The Political Economy of Oil-Exporting Countries

1. *Venezuela*
   Juan Carlos Boué

2. *Nigeria*
   Sarah Ahmad Khan

3. *Indonesia*
   Philip Barnes

4. *Libya*
   Judith Gurney

Libya
The Political Economy of Oil

Judith Gurney

Published by the Oxford University Press
for the Oxford Institute for Energy Studies
1996

Oxford University Press, Walton Street, Oxford OX2 6DP
Oxford New York
Athens Auckland Bangkok Bombay
Calcutta Cape Town Dar es Salaam Delhi
Florence Hong Kong Istanbul Karachi
Kuala Lumpur Madras Madrid Melbourne
Mexico City Nairobi Paris Singapore
Taipei Tokyo Toronto
and associated companies in Berlin Ibadan

Oxford is a trade mark of Oxford University Press

British Library Cataloguing in Publication Data
available

ISBN 0-19-730017-0

Cover design by Moss, Davies, Dandy, Turner Ltd.
Typeset by Philip Armstrong
Printed by Bookcraft, Avon

The Political Economy of Oil-Exporting Countries

Libya: The Political Economy of Oil is the fourth in a series of books on the major petroleum and gas exporting nations, most of them part of the developing world. These countries occupy a central position in the global economy given that oil, and increasingly natural gas are the energy sources most used in the world. Oil is also the most important primary commodity in international trade. At the same time the oil-exporting country despite progress in its efforts to diversify the economy still finds that its prospects are closely bound to the future of its oil.

Books in this series incorporate research work done at the Oxford Institute for Energy Studies. Their aim is to provide a broad description of the oil and gas sectors of the country concerned, highlighting those features which give each country a physiognomy of its own. The analysis is set in the context of history, economic policy and international relations. It also seeks to identify the specific challenges that the exporting country will face in the future in developing its wealth to the best advantage of the economy.

CONTENTS

TABLES

FIGURES

Major Libyan Oilfields

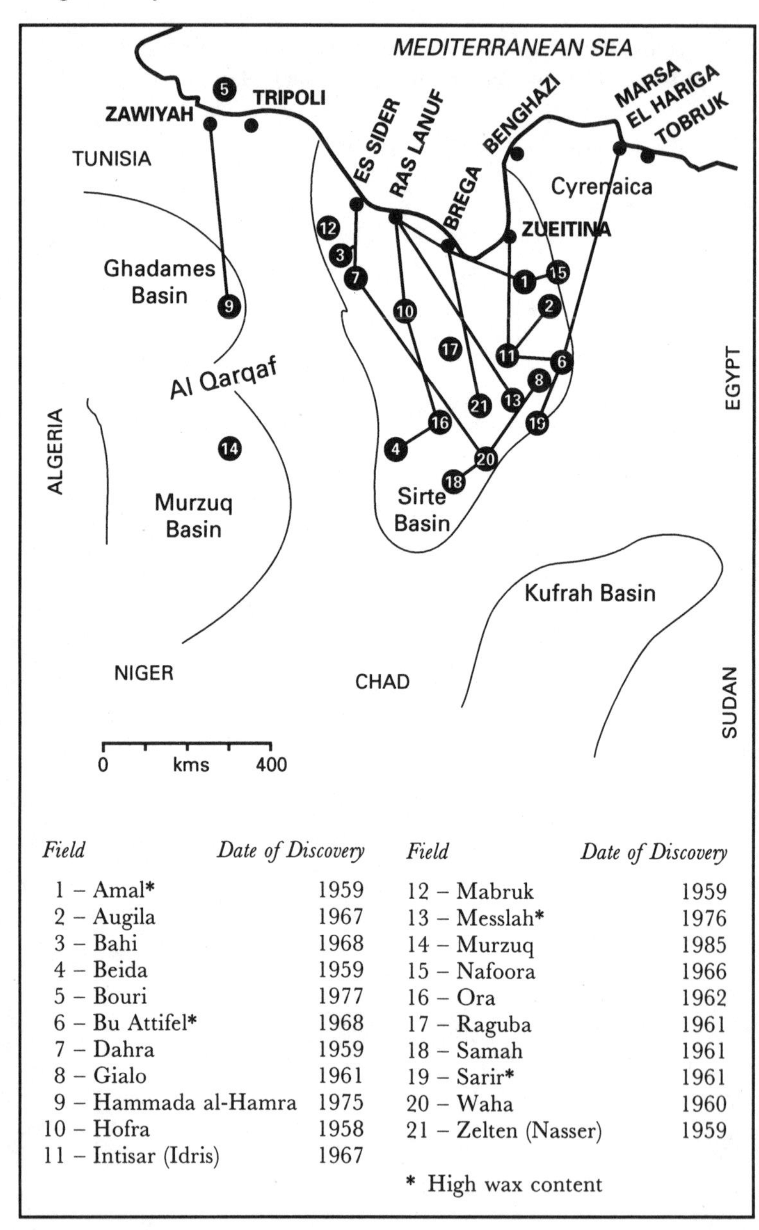

Field	*Date of Discovery*	*Field*	*Date of Discovery*
1 – Amal*	1959	12 – Mabruk	1959
2 – Augila	1967	13 – Messlah*	1976
3 – Bahi	1968	14 – Murzuq	1985
4 – Beida	1959	15 – Nafoora	1966
5 – Bouri	1977	16 – Ora	1962
6 – Bu Attifel*	1968	17 – Raguba	1961
7 – Dahra	1959	18 – Samah	1961
8 – Gialo	1961	19 – Sarir*	1961
9 – Hammada al-Hamra	1975	20 – Waha	1960
10 – Hofra	1958	21 – Zelten (Nasser)	1959
11 – Intisar (Idris)	1967		

* High wax content

1 INTRODUCTION

In many ways, Libya is an enigma. Three times in its relatively short history as an independent nation, it has aroused reactions of a far greater magnitude than would seem warranted to an objective observer, given its small population and lack of importance as a member of the international community. The first unusual response, when bountiful supplies of oil were discovered in the late 1950s beneath its desert, was one of enthusiasm and applause – despite the fact that there was then an abundance of undeveloped oil reserves elsewhere. The second, which by contrast was a reaction of shock and indignation, came in 1970 when Libya initiated a price bargaining round which led three years later to the destruction of the traditional system for setting oil prices by the major oil companies. The third response, one of rejection of Libya as an outcast, rogue nation, was occasioned by the radical political views of the Qaddafi government, including a lack of respect for the US government. It took longer to become full-blown. By the late 1980s, however, indignation had turned into condemnation and economic sanctions were imposed by the United Nations. Even the World Bank, which had invested so much interest, time and money into succouring the new nation in the 1950s, ceased to recognize the existence of Libya in its publications.

What is perhaps the most remarkable aspect of these widely expressed, volatile reactions is the low-key, matter-of-fact way in which the Libyan oil industry has been run from its very beginning under the monarchy, throughout the rule of the Revolutionary Command Council into the mid-1990s. The Libyan economy depends almost entirely on the production and sale of oil and there is a basic continuity in the management of these businesses which has resulted in a predominantly stable industry. The conservative nature of the government's oil policies over the years – with a few exceptions – is little recognized abroad and even less understood.

Although many in the industry still regard Libya as a producer of excellent quality oil with potentially significant undiscovered reserves, little outside attention has been focused lately on the structure of the country's oil industry and on the character of its oil policies. This study is an attempt to identify the internal and external factors which have shaped the industry and to analyse the implications for the future of Libya.

Viability Doubted at Independence

There were pathetically few expectations concerning the country which emerged from under the protection of the United Nations in December 1951. Libya was then one of the poorest nations in the world in terms of resources and income, with little potential for improvement. As Benjamin Higgins, the economist appointed by the UN to plan the country's economic and social development, wrote in his 1959 textbook of economic development:

> Libya combines within the borders of one country virtually all the obstacles to development that can be found anywhere: geographic, economic, political, sociological, technological. If Libya can be brought to a stage of sustained growth, there is hope for every country in the world.[1]

Descriptions of Libya at the time of independence make depressing reading. Except for a few clusters of settlements around one or two oases in the south-western province of Fezzan, its population of little over one million lived along the Mediterranean coast in a strip which varied in width from less than a mile to just over ten miles. With only approximately 1 per cent of its land area arable, and a further 4 per cent able to be used for pastoralism, the rest of the land area of 680,000 square miles was a grim desert where rain never fell. Its boundaries were arbitrary lines drawn up by European politicians and cartographers for their own purposes. Much of the north-eastern province of Cyrenaica had been ravaged by British, Italian and German desert armies during the war.

As for the Libyan people, some 90 per cent were illiterate and lived on a subsistence level with an estimated per capita income of $35 per annum. Of the handful of university graduates, none were doctors. The effect of a high rate of population growth was diminished by a high mortality rate, with one infant out of every two born alive dying within its first year. The people depended almost entirely for their livelihood on agriculture, whose expansion was severely limited. Barley and dates formed their traditional staple foods; the latter was considered to be one of the country's most important agricultural resources. Esparto grass, castor seed and scrap iron from the minefields were the country's main export goods. The accumulated capital of the vast bulk of the population consisted of livestock, tools, a little light equipment and housing that ranged from simple to primitive. Aid poured into the country from the UN, the USA, the UK, France and Italy, but to little avail. In fact, agriculture declined, partly because Libyans came to prefer imported gifts of wheat to indigenous barley.

There was an even bleaker picture in the industrial sector, with almost no electricity generation, no capital formation, no supply of skilled labour and no indigenous entrepreneurship. All the public utilities operated at a deficit. The banking system was primitive and the ownership of land and water rights was governed by tribal rules. There were no known mineral deposits large enough or rich enough to justify exploitation.

Competition for Oil

There was, however, the hope that oil existed and would be found. Few contemporary economists saw this possibility, even if it became a reality, as the means of salvation for Libya. The reason for their pessimism was that there was an abundance of available oil elsewhere. Reserves which had been discovered in the Middle East were not being developed as they were not needed, and there seemed little reason to believe that oil companies with ample undeveloped reserves in their portfolios would be interested in exploring the Libyan desert. The economists who despaired of the future of the new nation must have been amazed when Libya became the world's fourth largest exporter of crude oil, with one of the highest per capita incomes in the developing world, only a few years after oil was discovered in its desert hinterland.

Countries about to embark on oil exploration, and those considering inviting foreign oil companies back to aid in exploration, could do well to examine how Libya achieved this extraordinary result. It did so by recognizing the effectiveness of encouraging competition in the development of an oil industry. The government divided the country into a myriad of relatively small parcels which it offered as concessions for exploration and development at very little cost to any oil company willing to try its luck. Many did, as they felt that they had little to lose and wanted to preempt possible discoveries by others. The government further ensured competition by requiring companies to give back sizeable chunks of their concessions within a very few years; the returned acreage then went back into the pot for further disposal. This led companies to actively explore what they held, fearing that by failing to do so they might surrender acreage containing valuable undetected reserves – which did, in fact, happen on several occasions.

The decision to encourage a competitive oil industry might not have succeeded so well had it not been for two factors in Libya's favour: geology and location. The country had a fortunate prehistory when a warm, westward current flowed through the Tethys Sea, the precursor to the Mediterranean which stretched from the western Pacific to the Atlantic Ocean. This current dispersed tropical marine flora and fauna throughout the Middle East and along parts of the North African coast; these later decomposed to form crude oil and natural gas. Over millions of years, Libya's Sirte Basin, where most of the currently producing oilfields are found, was flooded by the Tethys. Over time, large deposits of oil and gas accumulated in sedimentary rocks beneath what became an arid desert. The oils in these Sirte Basin deposits were light in gravity and contained very little sulphur, a fact which became very important in an environmentally conscious world. Libyan oil came to be greatly valued by European refiners struggling to meet increasingly strong regulations regarding sulphur emissions.

Location was the second ace in Libya's pack. Historically, trade links had long existed between Libya and southern Europe. For years, caravans brought slaves, ivory, textiles and ostrich feathers from central Africa to Tripoli and Benghazi for shipment across the Mediterranean. In more modern times, it was oil that was traded to meet a booming demand in western Europe. Because Libya was situated so close to this market, its oil commanded a considerable advantage over Middle East oil in terms of cost of transport. It could reach European ports, especially those in the south, quickly and at low cost. Unlike Saudi Arabia, Libya did not depend on export pipelines which crossed the territory of other nations where they could be damaged or shut down. And unlike Iran and the Gulf oil producers, it did not have to ship its oil through the Suez Canal and meet increased transport costs if this waterway was unavailable. When the Suez Canal closed in 1967, *The Times* emphasized this point in an editorial in which it noted that 'It is difficult to overemphasize the importance of near-at-hand Libyan oil to the fuel economies of all the West European nations.'[2]

Realization of Strength

Libya confounded the world in 1970 by recognizing the value of its location at a time when there was brisk world demand for oil. The government opened a frontal attack on the oil companies operating in the country, demanding changes in arrangements concerning pricing and ownership. The aims were to increase its revenues and its control over the industry. It based its argument on the fact that although the prices at which the oil companies sold Libyan oil had increased, this had not been reflected in government taxation revenues which were based on an artificial 'posted price' set by the companies in 1961 and never altered. The immediate catalyst for the government's action was the closure of Tapline, the Saudi Arabian export pipeline which ran through Syria, as this resulted in vastly increased prices for oil in European markets.

The Libyan government's method for achieving its aim – reducing the production quotas for each company on a company-by-company basis at a time when oil was selling at a premium – was extremely effective. The capitulation of one company after another to demands for higher prices and thus increased levels of taxation precipitated a collapse in the entire price-setting mechanism for crude oil in Europe, the Middle East and Africa. Western countries' criticism was bitter and loud in the wake of Libya's actions and the subsequent Teheran and Tripoli agreements of February and March 1971 respectively. Blame was laid at Libya's door. The fact that the government overpriced its oil and was forced to make downward adjustments in order to regain market share was seen as just punishment for its actions. In fact, Libya subsequently adopted a moderate pricing stance and, in general, set its prices in line with those of other OPEC members. When the OPEC price system collapsed in 1985, it followed the practice of OPEC and non-OPEC producers by adopting a price formula based on the Brent market. Yet the fact that Libya has not been a price hawk for more than 20 years has hardly been noticed.

Embargo in Reverse

The vociferous condemnation of Libya in the late 1980s and early 1990s grew out of American anger over the extreme political pronouncements of the Qaddafi government and a belief that the latter was supporting terrorist movements. It was expressed first in a series of gradually more stringent US boycotts and eventually, in 1993, in UN sanctions. The US government had high regard for the effectiveness of economic sanctions despite historical experience to the contrary.[3] In fact, the first recorded example of the ineffectiveness of sanctions on their own occurred in 432 BC when the ancient Greeks issued a decree limiting the entry of Megara's products into Athenian markets in retaliation for Megara's attempted expropriation of territory and kidnapping of three women.

What was remarkable about the boycott and sanctions imposed by the West on Libya was that they involved the use of oil as a political weapon, as Arab nations had done in the Arab–Israeli wars of 1967 and 1973, and for which they were widely condemned in the Western world. Libya had been a staunch advocate of oil embargoes although it was the government of King Idris, not that of Colonel Qaddafi, which first ordered a ban on the sale of oil to countries believed to be supporting Israel during the Arab–Israeli war of 1967.

It is doubtful if the Libyan ban on the sale of oil to the United States, or the US ban on the purchase of Libyan oil, had much effect on the Libyan oil industry or economy. Economic sanctions rarely succeed in achieving much except when accompanied by warfare. What they tend to accomplish is the diversion of the direction of trade into other, and sometimes new, channels. The principal result of the US boycott of Libya has been the forging of strong ties between Libya with a number of continental European countries in upstream and downstream activities as well as in trade. This connection is publicly recognized. In late 1995, for instance, a ceremony was held in Libya to celebrate the production by the German company Wintershall of its 100 millionth barrel of Libyan crude oil. The Libyan oil minister took advantage of the occasion to emphasize that European oil companies had displaced American companies in the development of Libyan oil and gas.

Policy Constrictions

One of the cornerstones of Libyan oil policy has been a determination to avoid exclusive domination of the exploitation of its oil reserves by a few large multinational oil companies. It did not refuse these companies entrance; most were given the exploration acreage they requested in the early concession offerings, and most found important oil reserves in these. However, independent and European state oil companies also acquired good acreage and, in the end, most of the multinationals left for various reasons.[4] By contrast, many of the other companies stayed on and were joined by Far Eastern investors and other Europeans in the late 1980s and early 1990s.

At the heart of Libya's obsession with avoiding dominance by large foreign entities is perhaps its historical experience of invasions and occupation by Carthagians, Romans, Greeks, Persians, Egyptians, Byzantines, Turks and Italians. A fear of losing control has pervaded the range of Libya's relationships with other countries, with alliances frequently shifting and never allowed to become too strong.[5] Excessive caution may be the reason why Libya seems to shrink from opportunities to develop known oilfields and especially non-associated gas fields. It may be why the government is reluctant to open up certain unexplored areas to foreign companies despite repeated promises to do so over the years. It may also explain why the government hinted at new exploration and production-sharing agreements in the early 1990s but failed to set a date for these, and why, after expressing interest in arranging joint ventures, it postpones negotiations. The slowness with which the People's Committees of the Qaddafi government take decisions may be a result of Libya's reluctance to back intentions with deeds for fear of unseen consequences.

Libya's historical experience with invasions and occupations is at the root of another equally serious problem, lack of commercial know-how and self-confidence. When Libya gained independence after the Second World War it had no managers or middle class with experience in business and trade. During the Italian occupation, which extended over the period between the two world wars, Italians ran not only the government but all the public services and practically all commercial activities. The effects of lack of entrepreneurial experience are evident in the country's first marketing efforts and later, in its first downstream investments in Europe in 1985.

There are, however, apparent contradictions in Libyan policies. It is difficult, for instance, to reconcile the radical political thrust of the Qaddafi government with the basically conservative business practices of the Libyan National Oil Company. Strong commercial ties with Italy have continued despite the animosity engendered in Libya by its status as a former Fascist colony, Libya's downstream involvement in Europe has predominantly involved Italy and the majority of Libyan crude oil exports and all its liquefied natural gas in recent years have gone to Italy. In addition, the Italian oil company Agip plays a role second only to the country's national oil company in Libyan oil production and potential gas production.

Development of an Oil Industry

In the mid-1950s, when Libya became an independent nation and was pinning its hopes on finding oil reserves for its economic survival, the business of producing and marketing crude oil in the world had come a long way from its beginnings in the late nineteenth century. The industry had become more complicated and more mature. On the supply side, governments in countries where oil reserves had been found had acquired sufficient confidence in their sovereignty to demand a higher share of the gains from the production and sale of their oil. On the demand side, the postwar European market for oil was increasing at a dramatic pace, reflecting its substitution in stationary uses for indigenous coal and a greatly expanding transportation sector. Enlarged markets encouraged independent companies to enter the business, threatening the long-term control of the international industry by a handful of major oil companies.

Under these circumstances, the government of the newly independent nation felt able to lay down conditions under which oil companies could embark on exploration of oil in Libya. The process of drafting the 1955 Libyan Petroleum Law, which laid the framework for the subsequent development of the Libyan oil industry, is described in detail in Chapter 2. The government's goals were clear and it went about achieving these by taking two very significant initiatives. In the first place, it wanted to bring oil companies in to look for oil in Libya. In order to accomplish this at a time when abundant reserves of oil had been discovered elsewhere, it had to offer attractive financial terms.

In the second place, the government wanted to foster a competitive oil industry and to avoid a situation wherein oil companies took concessions and failed to explore or to produce these in order to suit their own worldwide reserve portfolios. Its 1955 Petroleum Law established a system that limited the size of concessions, mandated extensive relinquishment of acreage within a very few years and allowed not only large integrated oil companies but also smaller independents and state oil companies to become producers of Libyan oil.

The government gained by this competitive scene because it was able to pressure successfully individual companies for desired ends as no one company was essential to the Libyan oil industry and there were always others willing to take over a relinquished stake. The government lost, however, because competitive conditions encouraged companies to be too greedy and to produce too much too fast, often to the detriment of the oilfields.

Within a few years of the passage of its petroleum law, the government under King Idris realized that Libya contained prolific reserves of oil which were being feverishly developed. Even before exports were shipped out, it began to make efforts to increase its share of the profits; the success – and failure – of these efforts is described in Chapter 3. In negotiating with the companies that had been successful in locating oil, the government had the very great advantage in that the companies greatly prized Libyan oil which was then very cheap for them to produce and which commanded exceptional value in European markets because of its high quality and proximity.

But although the government was able to increase, at some times more than others, its profits from the production of Libyan oil, unlike many other OPEC members it never advocated complete nationalization of foreign companies. Both the Idris and Qaddafi governments accepted that they needed foreign companies to produce, and especially to market, Libyan oil and gas. Despite the confrontational nature of its relations with most foreign governments, Libya's industrial policy has been remarkably open, welcoming foreign companies, allowing them to operate within the country and observing accepted commercial practices with regard to contracts and payment. Its essentially carrot-and-stick relationship with foreign companies has been dictated primarily, in the final analysis, by hard-nose, economic considerations. The few instances of decisions affecting the industry which were taken for political reasons, such as the nationalization of BP and Hunt and the Amoseas partnership, were exceptional. Most attempts to increase government revenue and participation were done within at least a semblance of a legal framework, as were disputes arising over ownership of oil reserves. Border disagreements with Tunisia and Malta regarding sovereignty over offshore areas presumed to contain oil reserves, for instance, were finally settled by recourse to the International Court of Justice.

Most foreign companies continued their Libyan operations much as usual following the overthrow of the monarchy, at least for a few years. Some stayed on into the 1990s. As Chapter 4 indicates, Agip achieved an increasing presence in the country, with substantial production from its onshore and offshore acreage. The operations of Veba, Wintershall and OMV, although nowhere near to the scale of that of Agip, were still sufficiently profitable in the mid-1990s for these companies to continue their involvement in Libya.

Foreign companies, on the other hand, decreased their exploration efforts precipitously after the revolution as they felt uncertain of their future under the new regime and decided the wisest course was to concentrate on getting the most of what they had already found. By the early 1980s, the government realized that the companies that had remained in the country and the Libyan National Oil Company (NOC) were not capable of finding and developing new oil at a sufficiently high rate to replenish Libya's diminishing reserve base, especially with limited funds. In order to attract new companies to undertake exploration in Libya, and to interest those already in the country to expand their exploration efforts, it began to offer exploration and production-sharing agreements (EPSA). In the beginning, the terms of these EPSA offers, as they were called, were not attractive enough to encourage companies to become involved in exploration operations. In 1988, the government took note of this fact and its subsequent EPSA III round of offerings contained terms more favourable to foreign companies. Despite this, there had not been any substantial new discoveries in Libya by the end of 1995.

Nevertheless, companies continued to request Libyan acreage. What attracted them in the 1950s, and what has continued to do so, is the geological configuration of Libya, with its several sedimentary basins. As Chapter 5 indicates, virtually all of the original oil discoveries, many of which involved giant fields, were made in the Sirte Basin. Only in later years were finds made offshore Tripoli and in the western Murzuq and Ghadames Basins. Many geologists believe that there is more oil to be found in these basins and in the Sirte Basin as well, perhaps in stratigraphic reservoirs which are far more difficult to locate than anticline reservoirs. The exploration experiences of foreign oil companies in the first half of the 1990s in the Sirte Basin, however, were disappointing in this respect, possibly because they have largely limited their efforts to areas close to existing pipeline infrastructure.

The history of Libyan production is unusual in that, at least for a number of years, it did not reflect reserves or exploitation potential. It increased dramatically until 1970 when it peaked at 3.7 mb/d, more than twice the 1.5 mb/d produced annually between 1985 to 1995. It was reduced in the first instance as a result of government restrictions aimed partly at limiting damage to reservoirs which were believed to be over-producing and partly to get companies to consent to new taxation and participation agreements. OPEC quotas subsequently kept production at lower levels but there are doubts as to whether production, if not artificially controlled, could have increased subsequent to 1985, given the ageing of the producing fields, the absence of systematic application of enhanced oil recovery techniques and the low discovery rate of new reserves.

For a brief period in the early 1970s, the new revolutionary Libyan government took a very aggressive stand on price and, in doing so, served as the catalyst for a permanent change in the world oil pricing structure. By creating a new and different atmosphere regarding how prices were set, it brought the attention of the markets to the weakened monopolistic control over trade by the major oil companies, and changed the perception of all those involved regarding the factors which needed to be considered in calculating prices and how these factors could be manipulated to their advantage.

The grounds for the Libyan government's attack on pricing were laid in the 1960s, when it became apparent that the proximity of Libya to European markets meant substantial and significant transport advantages for its crudes over those from the Middle East. As Chapter 6 covering marketing and pricing in this decade indicates, the grounds for the aggressive stand taken by the Qaddafi government in the early 1970s lay in the effectiveness of OPEC members from the Gulf in persuading oil companies to agree to higher posted prices and therefore higher taxation levels. Flushed with its success in raising the level of prices for Libyan crudes, the government went too far and raised prices too high. Within a few months of the 1971 Tripoli Agreement, Libya had lost its cost advantage over Middle East oil in European markets and sales of its oil fell dramatically. Once the government realized its error it became much more conservative in setting prices. It came to adhere to OPEC pricing levels as a general rule and, when the OPEC pricing system collapsed, pegged its crudes to prices set by the Brent Market.

The government took a much more cautious approach to marketing than it did to pricing, partly because Libya had emerged as an independent nation after the Second World War with effectively no middle class and no experience in trade. It tentatively began to market its own oil in the last years of the Idris regime through the Libyan National Oil Company, which had come into being for this purpose in 1968. NOC's first moves in this direction were small and it followed the marketing pattern of the foreign oil companies which controlled all sales of Libyan oil through the 1960s. Like the independent oil companies which made up the Oasis partnership, and like Occidental, NOC first depended on third-party sales and then sought to secure its own wholesale and retail outlets in Europe. Eventually, as Chapter 7 shows, Libya's marketing policies deviated somewhat from this pattern as a result of the government's decision to use oil in barter and debt repayment agreements, particularly with the Soviet Union and eastern Europe.

The country's lack of entrepreneurial experience is evident in its broader downstream policies. It built refineries within the country without paying proper attention to its own internal product needs; neither of its two large refineries, the last completed in 1984, were designed to produce lighter distillates. This miscalculation resulted in a need to import gasoline and in using one of the country's highest quality crudes to produce fuel oil. The government's policy of relying on European refineries for products by providing these with regular deliveries of Libyan crude oil has never been profitable. Lack of internal refinery capacity for the production of gasoline has been exacerbated by the effect of US and UN sanctions which forbid the exportation to Libya of catalytic converters which would enable Libyan refineries to produce lighter distillates.

In addition, the criteria used by the government to purchase specific European refineries and retail outlets is difficult to justify. Decisions on these, as Chapter 8 indicates, appear to have been made on an opportunistic, *ad hoc* basis without the benefit of a calculated, long-term downstream programme based on supply and demand calculations. For the most part, Libyan downstream investment in Europe has concentrated on Italy. Libyans appear to be comfortable dealing with Italians, probably as a result of a long-standing historical connection between the two countries.

In 1983, Libya purchased shares in Tamoil, a company located in Italy which had a refinery at Cremona in Italy and later one at Collombey in Switzerland, as well as a network of retail outlets. The immediate impetus for this move undoubtedly lay in the worsening relations between the United States and Libya which increased the Libyan government's sense of urgency regarding the need for firm European connections. Two years earlier, the US government had requested all US citizens to leave Libya; a year later it banned the import of Libyan oil and placed restrictions on US exports to Libya. This was a period in which connections were widely believed to exist between terrorist activities and the Libyan government; it is possible that a secondary reason for Libyan investments in Europe was related to the need to have easy access to foreign exchange for the financing of underground activities. Whatever the initial impetus, once downstream investments in Europe began, Libya continued in this direction. It formed a company, Oilinvest, to control its overseas downstream investments which came to include the Holborn refinery in Germany. And although it nominally gave up majority control of Tamoil in 1991 in order to protect the organization from expected UN sanctions, it remained very much a driving force in the company after this date.

Libya's natural gas industry has also suffered from entrepreneurial inexperience, as Chapter 9 shows. Although Esso built what was then the world's largest LNG plant in Libya in the mid-1960s, the Libyan natural gas industry failed to reach its potential. For 25 years, production was effectively restricted to associated fields and facilities for liquefied natural gas were limited to the original Esso plant which had not been designed to extract LPGs, in which Libyan gas is unusually rich. Exports decreased over the years as a result of quarrels over price with Libya's two LNG customers, Italy and Spain. The discovery of several large non-associated gas fields in the 1980s revived interest in Libyan natural gas, but the wide geographic distribution of these fields poses serious difficulties for their commercial exploitation. Their development will undoubtedly require foreign investment and expertise which the government was apparently seeking in the mid-1990s.

Like many other desert countries in which oil has been found, Libya does not have physical resources other than hydrocarbons with which to sustain a viable economy. Before the discovery of oil, Libya was entirely dependent on agriculture for its survival although less than 1 per cent of its land area is considered fit for permanent cultivation. As Chapter 10 indicates, both the Idris and Qaddafi governments have sought to invest oil revenues in projects which would improve the country's basic economy; their success has been limited. Lots of money was poured into numerous agricultural schemes which essentially failed, with agricultural productivity continuing to decline. The Qaddafi government eventually focused on a major agricultural development project, the Great Man-Made River, designed to bring water from southern underground reservoirs to coastal areas. Completion of this project is expected by the late 1990s and should benefit the country for a number of years. The underground reservoirs, however, will eventually become depleted and agricultural production, in the long run, will become limited again. And, as noted earlier, investment in this project has starved the oil industry of essential funds. In the mid-1990s, it was reported that the government was reinvesting only 15 per cent of oil income into the oil industry although it was widely believed that 30 per cent was needed even to sustain oil production at its current levels.

There have been industrial investments as well, including petrochemical plants, fertilizer plants and an iron and steel plant. The economic benefits of these, particularly for employment, are not great. Overseas, Libya's main industrial investment was the purchase of shares in Fiat in 1977, which it sold in 1986. The immediate health of the Libyan economy depends on oil and gas production. Perhaps someday its impressive archaeological remains and beautiful coastline will bring tourism and thus another source of foreign exchange.

As the concluding chapter notes, however, two factors external to the functioning of the Libyan oil industry could have a strong impact on it and, as a consequence, on the Libyan economy. One of these is the continuance of sanctions over a long period and the other, unpredictable political changes. The risks involved with the eventuality of these external factors are difficult to quantify.

The importance of Libya in the 1960s stemmed from its largesse of excellent oil reserves and its proximity to European markets. This has not changed, and there is undoubtedly potential for the discovery of additional oil and gas reserves. Hopefully, an understanding of the evolution and thrust of Libyan oil policy will make it easier to assess the probable effect of changes that will, of necessity, occur.

Notes

1. Benjamin Higgins, *Economic Development, Principles, Problems and Policies* (1959), p. 37.
2. John Wright, *Libya: A Modern History* (1981), p. 233.
3. President Woodrow Wilson, in a speech in Indianapolis in 1919, is reported as saying that 'A nation that is boycotted is a nation that is in sight of surrender.'
4. The first to go was British Petroleum, which was nationalized in 1971 in a political gesture as was the US Amoseas partnership composed of Texaco and Socal, and Arco's share of Esso Sirte, in 1974. Several US companies went of their own accord and those that remained were ordered to leave by their government in 1986. Shell finally abandoned its last, unsuccessful claim in 1991.
5. A policy of risk aversion has adversely affected the oil industry obliquely. The government's decision to finance the Great Man-Made River project entirely from government revenues, for instance, has had the disastrous effect of starving the oil industry of essential investment funds for a number of years.

2 A PETROLEUM LAW WITH UNUSUAL FEATURES

One of the most important factors shaping Libyan oil policy, and the subsequent development of the country's oil industry, is the date when it all began. The Libyan Petroleum Law was promulgated in 1955, at a time when oil companies were shifting their emphasis from the production of decreasing reserves in the United States to overseas locations. Demand for oil was increasing dramatically in Europe, despite the lack of known reserves in this region, reflecting the substitution of oil for indigenous coal. Expanding markets encouraged independent oil companies to enter the business, threatening the long-term control of the industry by the few 'majors'. On all fronts, the players were becoming more numerous and perhaps more sophisticated.

At that time, however, Libya could scarcely be described as having the potential to become a viable independent nation. Its economy was almost entirely agricultural although 95 per cent of the land area was desert and only about 1 per cent was considered fit for permanent cultivation. Prewar, Libya had been a colony of Italy, its agriculture subsidized in order to aid the settlement of Italian farmers in Libya. Its few small factories producing agricultural foodstuffs, tobacco, building materials and metal products had been owned and run mostly by Italians and many had ceased to operate by 1945. In other words, while the Fascist administration had created a basic infrastructure this was for the benefit of the Italian settlers brought into the country and the native Libyans had suffered in consequence. Capital formation was effectively non-existent and there was virtually no Libyan skilled labour or entrepreneurship. On the plus side, it had remarkable ruins of ancient civilizations and a beautiful coastline.

If oil were to be found, Libya's chances for survival would increase dramatically, for it was well positioned to supply the southern shores of France and Italy with their energy needs. The challenge for those formulating the country's petroleum law was to ensure that Libya, as well as the oil companies involved, gained if oil was discovered. In order to understand what evolved, it is necessary to position the oil era which followed the promulgation of a petroleum law in the political and economic context of this extremely poor and undeveloped country.[1]

An Independent Nation

Libya achieved independence in December 1951, after almost seven years of legal wrangling over its status following the end of the war. The first step in gaining independence was the Italian peace treaty, signed in February 1947, in which Italy renounced its claim to the colony that it had acquired from Turkey in the First World War. According to the terms of this treaty, the Four-Power Council of Foreign Ministers of Britain, France, the United States and the Soviet Union were to decide on the future status of Libya. If they failed to reach an agreement within one year of the date of the treaty's coming into effect, the question would be referred to the General Assembly of the United Nations. The views of these foreign ministers, however, instead of reaching a consensus grew farther and farther apart over the months that followed, reflecting the intensification of the Cold War.

As East/West distrust increased, all four nations considered the geographical location of Libya to have paramount strategic importance. The United States and Britain feared Soviet expansion in the Middle East, their suspicions having been fuelled by the reluctance of the Soviet Union to withdraw its troops from Iran at the end of the war. They felt it was important to secure the North African coast in order to protect southern Europe. The British Military Administration's determination to retain its presence in the eastern Libyan province of Cyrenaica was intensified by the growth of anti-British feeling in Egypt. US military planners were resolved to retain the Libyan Wheelus airfield which they had acquired during the war as a critical link in a planned network of American military installations encircling the Communist bloc with air strike and air defence capabilities. The French intended to hold on to their military and civilian settlements in the south-west desert province of Fezzan, fearing that granting independence to the Arabs in Libya would have an unsettling effect on the French empire in North Africa, particularly in Algeria and Tunisia. The Soviet Union, quite simply, wanted a foothold in North Africa and Libya seemed the most likely place to get one.

Faced with deadlock, the four foreign ministers referred the matter to the General Assembly of the United Nations in September 1948, in accordance with the terms of the Italian peace treaty. Italy made a bid to reassume, in part, its former role in Libya by securing for itself the trusteeship of the province of Tripolitania. In a move designed to block a Soviet proposal, first voiced in 1945, for a similar Tripolitania trusteeship, many countries supported the Italian suggestion. Ernest Bevin, the British foreign secretary, and Count Carlo Sforza, the Italian foreign minister, submitted a plan for the three Libyan provinces to be ruled separately, with Cyrenaica placed under British trusteeship, Fezzan under French trusteeship, and Tripolitania under Italian trusteeship.

When the so-called Bevin–Sforza plan was submitted to the UN General Assembly for approval in May 1949, the Italian proposal for trusteeship of Tripolitania was defeated by the single vote of Emile Saint-Lot, the delegate from the republic of Haiti who was later made an honorary Libyan citizen. This led to the subsequent defeat of the plan.

Seven months later, in November 1949, the UN General Assembly voted that Libya should become a united, independent nation in a year's time. The United States favoured this decision on the grounds that an independent Libya could enter freely into agreements with the Western powers regarding military bases on its territory, an activity which would have been forbidden under a United Nations trusteeship.[2] Italy withdrew its bid for Tripolitania in the hope that this gesture would occasion goodwill on the part of future Libyan governments.

In the year that followed this decision, the UN assistant secretary, Adrian Pelt, served as a UN commissioner in Libya, preparing the grounds for independence. By overseeing the drafting of a constitution by representatives from the three provinces the UN, in deference to contemporary international law, hoped to ensure the sovereign rights of this former Italian colony.

Dependence on Foreign Aid

When it was officially declared independent in December 1951, Libya had a population of just over one million Arabic-speaking Muslims. The Council of Foreign Ministers, the United Nations and the International Bank for Reconstruction and Development (IBRD) all sent commissions to report on the new nation's prospects, and these painted a grim picture of poverty.[3] As noted above, over 90 per cent of the population was illiterate, and the new country was one of the poorest states in Africa. Banks had been closed since the middle of the war. There was no trade with Italy and only a little with Egypt and Tunisia. The eastern province of Cyrenaica had suffered severe war damage and agriculture, on which the economy entirely depended, was in a poor state. The UN mission cited poor soil, frequent droughts and lack of modern methods and machinery as hindering crop production while the IBRD group singled out the system of tribal ownership of land and water rights as at fault.

The only remnants of industrial activity were in Tripolitania where a few small cottage industries, mostly run by Italians, were involved in agriculture-related processing, such as flour milling, olive oil refining, date packing, and tobacco and salt manufacture. There was also some production of textiles such as carpets and blankets, footwear, and building materials, including tiles, brick products and limestone blocks. The only known mineral resource deposits were some iron ore in the Fezzan, some gypsum near Tripoli, some potash in the Sirte desert and occasional traces here and there of sulphur, manganese, lignite, alum and sodium carbonate.

All the mission reports agreed that there was no question of economic self-sufficiency. Income from exports, which consisted mainly of esparto grass used in the manufacture of high-grade paper such as bank notes, citrus fruits, olive oil, livestock, hides and skins, ostrich feathers and, in years of good harvests, barley, could not begin to cover the cost of essential imports. A valuable export in the immediate postwar years, scrap metal left over from wartime, was a highly depletable resource.

The missions pointed out that even during the prewar era, despite considerable Italian investment, the balance of trade had been in deficit, the public utilities and railways had run at a loss, and few of the Italian agricultural schemes, such as wheat and tobacco farms, had made a profit. The new nation which emerged in 1951 would require considerable sums of foreign capital to survive, to repair wartime damage and to develop a viable economy.

The United States, Britain and to a lesser extent, France, provided financial aid, fearing the encroachment of communism in Libya if they failed to help alleviate poverty. The United Nations supported technical assistance programmes. All this aid, however, was insufficient to cover even the deficits in the balance of trade which rose from £3.6 million in 1950 to £8.8 million in 1951. More money was needed to raise the country from its status as one of the poorest in the postwar world. This came from the United States and Britain in return for agreements that they could continue to occupy military bases in Libya.

Hopes for Oil

Expectations that Libya had undiscovered crude oil reserves were based on the existence of a belt of sedimentary rocks stretching across North Africa. In 1938, an Italian geologist, Ardito Desio, reported finds to the Italian government of oil traces in water wells in several coastal locations and prepared a geological map of Italian North Africa. In 1940, the government instructed Aziendi Generale Italiana Petroliche (Agip) to explore the Sirte Basin. This exploration, severely hampered by desert conditions and inadequate facilities, was abandoned after the outbreak of hostilities.

Postwar exploration in Algeria had located oil and gas fields by the early 1950s and Libya was judged to have many of the same geological features as Algeria. Several major oil companies sent survey teams to Libya in the postwar years, before the 1955 Petroleum Law was in place, although they had no legal basis on which to start exploration. Standard Oil of New Jersey (Esso Standard), for instance, made excursions into Libyan territory and came to the conclusion that the chances of discovering oil in commercial quantities in Libya were good. Company internal documents of 1947 noted that what they had learned in Libya was 'useful'.[4]

D'Arcy Exploration Co. Ltd. (British Petroleum) also undertook aerial and ground surveying at this time. A 1953 letter from N.A. Gass, a BP executive, to Guy Moore, who had suggested exploration along the Libyan border with Sudan, noted that:

> We have, in fact, been studying the oil possibilities in Libya for some time. Several of our geological parties have already visited the country, and we have been in touch with the Libyan Government in connexion with the possible acquisition of exploration rights. Unfortunately, however, there is as yet no Libyan legislation under which exploration permits or licences can be issued. The necessary regulations are, we understand, now being drafted, but until they are actually enacted, there is no further action which we can take meanwhile.[5]

Royal Dutch Shell also was active in early exploration. A geological conference held in London in 1952 reported, with regard to Libya, that 'a certain amount of gravity work had been done by Shell; this showed a regional low over Sirte which ties in with what is known in Sicily, Malta, etc.'[6]

Requirements for a Petroleum Law

Laying down terms under which foreign oil companies could look for and produce Libyan oil reserves was a protracted process. The first step, to invite companies to come and look at the possibilities, was relatively simple and was accomplished by a legislative act, the Minerals Law of 1953, published in the *Official Gazette* of 18 September of that year. This stipulated that the Libyan state was the owner of all minerals, including hydrocarbons, in the subsoil and laid down the conditions under which companies could obtain permits to carry out survey work for petroleum resources in requested areas. Such permits, however, did not allow drilling or seismic operations and did not guarantee a permit holder a future claim for a production licence in any area which it had explored.

Nine companies were granted one-year survey permits in late 1953 and 1954: Esso Standard (Libya), an affiliate of the Standard Oil Company of New Jersey; Anglo-Saxon Petroleum Company, an affiliate of Royal Dutch Shell; D'Arcy Exploration Company (Africa) Ltd., an affiliate of British Petroleum; Mobil Oil of Canada, an affiliate of the Mobil Oil Company; Compagnie Française des Pétroles (Total), American Overseas (Amoseas), acting as operator for California Asiatic Oil Company and Texaco Overseas Petroleum Company; Oasis Oil Company, an affiliate of Ohio Oil Company later named Marathon Petroleum (Libya), acting as operator for Amerada Petroleum Company and Continental Oil Company; Nelson Bunker Hunt, and the Libyan American Oil Company (Arco). Although permits were due to expire in one year, D'Arcy, at least, obtained an extension until well in 1955.[7]

In the meantime, the Libyan government and its advisers set to work drafting a comprehensive petroleum law outlining the terms under which oil companies could acquire concessions for exploration and production. Oil company prewar concession agreements with oil-producing countries in the Middle East had been struck when these countries were undeveloped and in many instances not fully independent. Iraq, for instance, was a British mandate and Kuwait, Bahrain and Qatar were in close treaty relationship with Britain. Concession agreements often granted oil companies long-term exclusive rights to exploration and production of the entire country; in instances where a whole country was not included, they covered a major region which, either because of its geological formation or because of its political character, consisted of a well-defined territorial entity.

Oil companies could no longer expect to get these types of blanket concession agreements after the Second World War. But while they had to give greater consideration to the sovereignty of governments of countries in which oil might be found, they had wider access to possible oil-bearing sites. The prewar 'Red Line Agreement' delineating the spheres of influence of oil companies in the Middle East was no longer recognized. Under its terms, US companies had agreed to refrain from activities in territories falling within the borders of the former Turkish Empire except in conjunction with the partners in Iraq Petroleum Company. Freed from this restraint, Standard Oil of New Jersey (Esso) and Socony (Mobil) had begun operations in Saudi Arabia and other US oil companies had expanded their involvement in the Middle East.

Exploration during the late 1940s and early 1950s by American companies and others in Saudi Arabia, Kuwait, Iraq, the Neutral Zone and the Trucial Coast proved very successful and estimated Middle East reserves increased, on average, more than 14 per cent annually between 1949 and 1953. This growth in supply allowed oil companies to meet the postwar surge in demand for oil which, in Europe alone, was then increasing at an annual rate of 13 per cent. With ample current and expected reserves in Middle East countries where they were established, the companies did not need to explore areas where prospects for discoveries were low or unknown, or where concession terms were poor.

The drafters of the Libyan Petroleum Law were aware that they had to offer attractive terms to persuade companies that exploration and production in the Libyan desert would be worth their while. In doing so, however, they had to tread carefully. In the first place, they had to contend with the risk that oil companies would lease acreage in Libya which they did not intend to explore until it suited them. The wide-ranging, exclusive rights which companies had been able to acquire in early Middle East concession agreements enabled them to engage in preemptive behaviour when circumstances seemed to require it. This was not a unique characteristic of the oil industry in the Middle East. Companies and individual prospectors in the United States had a long history of leasing areas which they had no intention of developing – at least in the short term – merely in order to prevent other companies from acquiring them. In 1933, for instance, US oil companies had 93.5 million acres of land under lease, but only 9.9 million of these were listed as having proved reserves.[8] The Libyan government did not want this type of preemption, which might mean that its only hope for prosperity was indefinitely postponed.

A second closely allied danger was that an oil company which found oil in Libya would limit production to suit its own needs. Oil companies involved in the Middle East were known to have adjusted the production levels of their different holdings for various reasons. Production restraint in Saudi Arabia during the Second World War, for instance, was largely the result of the lack of refinery capacity in that country, given that there was a working refinery at Abadan in Iran. The reluctance to develop the oil-bearing region of Qatar before the war, on the other hand, was the result of the excess supply of oil on world markets during the 1930s. Although the Anglo-Iranian Oil Company (AIOC), and later the Iraq Petroleum Company (IPC), had exclusive exploration rights in Qatar as early as 1932, drilling did not start until just before the war and production was delayed even later. Proof of deliberate production restraint in Qatar was submitted to a US Senate committee investigating the activities of US oil companies in the Middle East. A memo written by an oil official in 1941 noted that 'as there is excess of petroleum products available from AIOC and Cal-Tex in Persian Gulf, it is obvious productive wells in Qatar will not be expedited at the present time'.[9]

IPC had also restrained prewar production in Iraq. Testimony to the same Senate committee described IPC as having:

> employed a variety of methods to retard developments in Iraq and prolong the period before the entry of Iraq oil into world markets ... Restrictive practices were continued even after a pipeline was completed ... in 1935, IPC's production was shut back several hundred thousand tons.[10]

The committee was told that IPC was identified with a policy of restrictive production to such an extent that it hesitated to bid for new concessions on the grounds that its 'object in obtaining fresh oil territory would not be associated with any irrepressible urge for intensive exploitation'.

Drafters of the Libyan Petroleum Law wanted to ensure that once companies were given acreage and found oil, they produced this oil at a level which would generate an adequate income for the Libyan government.

Not all the postwar changes, however, represented risks to the development of a Libyan oil industry. The significant changes in the type of financial arrangements between oil companies and Middle Eastern governments which were occurring in the early 1950s were considerably to the benefit of these governments, and Libya was in a position to demand more revenue from the production of oil than it would have been even five years earlier, if it wished to do so.

Up until 1950, Iran, Iraq, Saudi Arabia and Kuwait received their major payments from oil companies holding concessions in their territories in the form of royalties based on the tonnage of oil produced. The first royalty fee incorporated in a Middle East concession agreement was put at 4 gold shillings per ton of oil produced, an amount considered to be one-eighth, or 12 per cent, of the value of Middle East oil at that time. The use of a 12 per cent share came from the United States. In the early days of the oil industry in Pennsylvania it was estimated that a quarter of the market price of petroleum was profit. Although 12 per cent may have once represented an equal sharing of profits between land owner and producer in Pennsylvania, it translated into a far different profit-sharing arrangement in the Middle East, where production costs in oilfields were much lower.

The governments of the Middle East oil-producing states with 12 per cent royalty agreements were naturally concerned with upstream production levels rather than with downstream prices and profits. For many years, they were largely unaware of the cost of producing oil or of the price at which it was sold, the latter being essentially a transfer price between various subsidiaries of vertically-integrated oil companies. After the Second World War, however, these governments became increasingly aware of oil company profits, which were then increasing substantially, and began to demand a share of these.

Venezuela was the first oil-producing country to make a firm move in this direction. In 1948, it levied a tax of 50 per cent on the profits of oil companies working in that country. Following the Venezuelan example, Saudi Arabia, aware of its military security importance to the US government in the Cold War, insisted that it should have a 50 per cent share of Aramco's profits. George C. McGhee, a US Assistant Secretary of State, indicated to Aramco that the government wished it to accede to the Saudi demand. Aramco was willing to comply only if it was allowed to offset the additional funds paid to the Saudi government against its US income tax liability. Negotiations between the government, the company and the US Treasury led to a solution whereby Saudi Arabia's 50 per cent share of Aramco's profits was deemed to be an income tax eligible for foreign tax credit under existing US law. Negotiations for revised concession agreements in other Middle East producing countries followed, focusing on combining royalty payments with an income tax in a manner calculated to give governments a half share of the profits and the US companies involved, foreign tax credit.

The taxation systems which emerged in Middle East oil-producing countries after these negotiations resulted in government revenues which were calculated on posted prices – the prices at which a posting company theoretically was willing to sell its oil to third-party buyers at point of export. Posted prices were higher than realized prices at the time that the Libyan government was drafting its petroleum law. By specifying that taxes would be levied on realized prices rather than posted prices, Libya could give foreign companies that found oil an opportunity to achieve higher profits. As realized prices diverged more and more from posted prices after 1957, this inducement became increasingly attractive to companies.

The negotiations over the tax issue in Saudi Arabia, and those that occurred later when the nationalization of Anglo-Iranian was threatened, indicated that Western governments were inclined to intervene in disputes between oil companies and foreign governments when they felt that this was necessary. In later years, such intervention played a large role in the Libyan oil industry. The US government, for political reasons, came to play an exceptionally high profile role in the relationship between US oil companies and the Libyan government.

A further important change in the oil industry globally after the Second World War which the drafters of the Libyan Petroleum Law needed to take into consideration was the type of players involved. Although the eight major oil companies still remained in control of oil production and marketing, there were now increasing numbers of smaller, mostly American, independent oil companies in the industry. Many of these independents were actively seeking reserves outside the United States; in 1945, only 20 US companies were engaged in exploration abroad and by 1958 there were some 220.[11] This overseas focus was partly the result of the decline of the United States after the Second World War as the major source of petroleum supply. In 1938, US crude oil production was 59 per cent of world production; by 1949 this had fallen to 51 per cent and by 1955 to 42 per cent. Moreover, domestic crude production was proving unable to satisfy demand and the United

States became a net importer of oil after 1948. US net imports as a percentage of consumption increased from 5.5 per cent in 1949 to 10.4 per cent in 1955. (They reached 16.7 per cent in 1959 when the US government introduced an import quota.)[12]

Another reason why the independents were looking abroad was to escape domestic oil production regulations. The principal US regulatory body, the Texas Railroad Commission, decreed limits for crude oil output per well in accordance with its estimate of reasonable demand and maximum rate of efficient production. Similar regulatory bodies which set output rates in other oil-producing states tended to follow the limits set in Texas. Independent oil companies, in particular, found such production restrictions a severe restraint on growth.

There was undoubtedly an opportunity for Libya to attract these independent US companies, with their hunger for reserves, by offering attractive terms in its petroleum law. In the long run, this turned out to be very important.

Laws Drafted in Turkey, Egypt and Israel

Libya was not the only country preparing a petroleum law at this time, although it had the considerable advantage of being able to offer more hope for the discovery of oil reserves than most. Three Middle East/North African countries enacted petroleum laws in the early 1950s.[13] The similarities in the terms of these laws, which found echoes in the 1955 Libyan law, reflected the effect of the industry changes discussed above. They were also the result of the advice of several American lawyers and geologists hired by the governments in these countries as consultants during the drafting process. There is no evidence that any of these same Americans were directly involved in advising on the content of the Libyan law. There is no doubt, however, that those preparing the Libyan law were aware of what was being done in the countries noted below.

Turkey. In 1952, the Turkish government decided to privatize the country's oil industry. Up to this time, only the state oil company was permitted to carry out oil exploration and development although it was often assisted in technical operations by American firms on short-term contracts. Very few petroleum reserves had been found and produced, partly because of a lack of investment, and increasing imports of oil were creating serious financial problems in Turkey's balance of payments. Shortages of fuel were hampering ambitious postwar industrial development plans.

George McGhee, a former Texas independent oilman and geologist who served as American Ambassador to Turkey from 1951 to 1953, strongly advised the government to privatize the industry.[14] This was the same George McGhee who, as US Assistant Secretary of State for Near East Affairs, had brokered the tax agreement between Saudi Arabia and Aramco and who had played a key role during the nationalization crisis in Iran. The Turkish government, heeding McGhee's advice, announced that it was abolishing the state monopoly and preparing a petroleum law outlining the terms for private company participation in the Turkish oil industry. It was McGhee, apparently, who suggested to the Turkish government that they retain Max Ball, an American petroleum consultant from Washington DC, to draft the new petroleum law. Max Ball was a petroleum geologist educated at the Colorado School of Mines and a former director of the US Interior Department's Oil and Gas Division which had succeeded the wartime US Petroleum Administration. He had also worked for Shell and other oil companies and was familiar with petroleum prospects in the Middle East.[15] There was criticism within Turkey on the propriety of hiring a geologist to draft a law which might be slanted to his own interest. The Turkish minister for exploitation of national resources argued that Ball, along with other foreign experts, 'had come to study how best Turkish oil could be extracted, and had not demanded any concessions as their perquisites'.[16]

Max Ball, in turn, retained Elmer Batzell, a lawyer from the Washington DC firm of Meyers and Batzell, who had served as assistant chief counsel for the wartime US Petroleum Administration, to prepare a draft law.[17] Together, Ball and Batzell persuaded the Turkish government to forego high rents, royalties and other entrance fees, at least initially, on the grounds that the benefits of abundant petroleum discoveries would be far greater than anything the government could derive from heavy taxation of oil companies, at least in the early years. They believed that petroleum laws should encourage competition in the oil industry, particularly by American firms.[18]

While the law was being drafted and discussed, the Turkish government authorized several foreign oil companies to undertake preliminary reconnaissance work. Exploration permits were granted to ten companies, including four which also had exploration permits in Libya at this time – Esso Standard, Royal Dutch Shell, Socony-Vacuum (Mobil), and Amoseas (an affiliate of Standard Oil of California and Texaco). Thus at least some, if not the majority, of the companies consulted during the formulation of the Libyan Petroleum Law were aware of the law under consideration in Turkey which, incidentally, was originally drafted in English.[19] D'Arcy Exploration (BP) advisers had studied it carefully. A D'Arcy memorandum of 5 August, 1954, commenting on an early draft of the Libyan law, noted: 'Section 2 ... of the Turkish Petroleum Law we believe might be cited as a good example of a concrete definition of a Petroleum Administration's or Commission's organization and functions.'[20]

In March 1954, the Turkish National Assembly approved the new petroleum law designed to promote rapid development of oil production in Turkey by encouraging competition and offering inducements to foreign companies. Several of its terms were unusual and similar to ones later included in the Libyan petroleum law, sometimes in a slightly different form. These included specifications that:

i. The country was divided into seven zones for the purpose of allotting concessions.
ii. Production leases could not exceed 500,000 acres in any of the seven districts.
iii. An independent Petroleum Administration was established to handle the implementation of the law and to decide on the granting, assignment, renewal, surrender and revocation of concessions.
iv. In calculating profits, the companies were to be granted all the allowances which the US government granted to oil producers in the United States for tax purposes. This included an annual depletion allowance of 27.5 per cent of gross income less rentals and royalties.
v. After a company recovered its initial investment, profits were

> to be subject to a special surtax which would give the Turkish government 50 per cent of the net profits, including royalties. Royalty, rents, income and other local taxes were not to exceed 50 per cent of the net profits calculation before taxation but after the depletion allowance.[21]

Egypt. The Egyptian government made several changes to its petroleum laws in the early 1950s. Although oil began to be produced in commercial quantities in Egypt in 1912, the first petroleum legislation was not enacted until 1948. Prior to that time, there were a number of government regulations regarding licences to explore and produce oil. A number of companies had acquired petroleum leases between the wars, but few had been successful in finding oil; the notable exception was Anglo-Egyptian Oilfields, Ltd., a Royal Dutch Shell subsidiary, which held leases on the main producing fields until the late 1930s. Many applications for exploration licences were made following the discovery of oil in Saudi Arabia and the concurrent discovery of a major Egyptian oilfield south-east of Suez. In addition to Shell, Esso, Caltex (Standard Oil of California and Texaco) and Socony-Vacuum (Mobil) were among the major companies interested in Egypt prior to the Second World War.

In 1948, the first comprehensive Egyptian petroleum law and subsequent regulations increased the amount of royalty payments on oil production and restricted the freedom of foreign companies to operate in Egypt by stipulating that only Egyptian companies could obtain oil-mining licences. Most companies refused to accept these changes and exploration was severely curtailed. Decreasing production in the late 1940s was a serious blow to the country which was already importing large quantities of petroleum products to meet domestic demand. Continued exploration was critical, given that many of the early fields had been exhausted.

In 1950 and 1953, the government drew up new legislation permitting foreigners to hold up to 51 per cent of the shares in Egyptian companies and outlining more favourable conditions for foreign oil companies. In 1954, a contract was concluded between the government and the Conorada Petroleum Corporation, a US joint venture consisting of the Ohio Company (later named Marathon), the Continental Oil Company and the

Amerada Petroleum Company for exploration and production of petroleum reserves in vast areas of the Western Desert. This contract, involving a concession lease for 30 years renewable for another 30 years, was drawn up in accordance with the new law. It contained the requirement that Conorada must relinquish 25 per cent of the territory under lease at the end of the third year and another 25 per cent at the end of the sixth year. At the end of the twelfth year, it could only retain less than a fifth of the original territory.[22] Severe relinquishment requirements of this type for similar desert areas were incorporated in the Libyan petroleum law.

Israel. Israel was importing large quantities of oil in the late 1940s, a fact which exacerbated its critical foreign exchange position. It wanted to attract foreign oil companies to undertake exploration for oil reserves, but faced a number of difficulties. Companies involved in the Arab Middle East were extremely cautious about getting involved in Israel. In addition, there was little reason to believe that extensive oil reserves existed there, given the results of exploration done in Palestine before the war.

The best course of action, in the view of the Israeli government, was to draw up a petroleum law which would attract sufficient numbers of foreign oil companies to ensure competition in exploration. It intended to create a situation in which petroleum reserves 'would be found as quickly as possible' and produced 'in accordance with well-established business and operational principles existing in the oil industry'. The oil industry should be 'opened to free and competitive exercise' and private capital should be given 'valid opportunity for development'. The 'alternative policy of awarding blanket concessions to an oil company which would operate on the basis of an exclusive national monopoly was considered and rejected'.[23]

In August 1952, the government passed a petroleum law which was largely the result of a draft submitted, at the government's request, by Max Ball, the same American petroleum geologist who had been retained by the Turkish government. It would appear that Max Ball adapted the Turkish draft law prepared by Elmer Batzell to fit Israel's requirements, although he did not acknowledge the latter's role.[24] The draft was then circulated among lawyers and businessmen in the United States for comment. The Israeli Petroleum Commissioner noted that 'with the help of American experts, a Petroleum Law and Regulations were drafted offering the operators a favourable economic climate with operating conditions which are not inferior to those prevalent in their home countries'.[25] But whereas in Turkey government officials could assure critics that the interests of the foreign experts it retained were purely advisory on legal matters, this was not the case in Israel. Two years earlier, Max Ball and his son Douglas had been hired by the Israeli government to do an extensive geological survey of the country. Douglas Ball, like his father, was a petroleum geologist educated at the Colorado School of Mines. The results of this survey, published in the United States in 1953, suggested the presence of at least some, if not considerable, petroleum reserves.[26]

Not surprisingly, given the involvement of the same consultants, the Israeli law resembled the Turkish law. The country was divided into four petroleum districts corresponding to its main geographical areas and the acreage of every concession lot was relatively small. A company was limited in the number of concessions which it could hold in any district. Royalty was set at 12 per cent and companies were subject to Israeli taxation at a rate of 50 per cent. A Petroleum Board was established to 'advise in all those matters which, it was felt, contained within them a possibility of serious prejudice to an operator'.[27]

The 1955 Libyan Petroleum Law

There is no question but that the Libyan government had foreign consultants with oil industry and legal experience serving as advisors regarding the contents of its petroleum law. Abdul Amir Q. Kubbah, who served as a petroleum economist in the Libyan Ministry of Petroleum from 1962 to 1964, mentions the presence of A. Hogenhuis, who was retired from Royal Dutch Shell, and of C. Andrews from Britain.[28] Another source mentions assistance given by Messrs Dale and Pitt-Hardacre, the latter described as a British adviser.[29] There is no evidence that any US advisers were directly involved, but Kubbah relates that the preliminary draft drew heavily on the American-drafted Turkish law.[30]

The Libyan government submitted the original draft of its law, along with accompanying schedules, to at least some of the oil companies which had expressed sufficient interest to apply for exploration permits.[31] The document reiterated that all natural deposits of petroleum were the property of the Libyan state and gave provincial authorities the right to continue to grant exploration permits in areas not included in existing permits or in pending concession applications. Some of its terms resembled those which had recently been agreed in Turkey, Egypt and Israel. These included:

i. The establishment of a Petroleum Commission to administer the law and to implement the government's rules and regulations.
ii. Limitation of the size of a single concession to no more than 75,000 square kilometres.
iii. Limitation of concession holdings by a single oil company to a total of no more than 225,000 square kilometres, and to no more than 75,000 square kilometres in one province.
iv. A requirement that a company must relinquish 25 per cent of its concession in the fifth year, another 25 per cent in the eighth year, and all of the concession in excess of 10,000 square kilometres in the tenth year.
v. Specification of a royalty of 12½ per cent to be levied on the established market price f.o.b at a Libyan seaboard terminal.[32]
vi. Calculation of the total sum of rents, royalties, income taxes and other levies paid to the federal and provincial governments at a level equal to 50 per cent of an oil company's profits.

In addition, the draft required that companies must commence operations within a set time period and must expend specified sums on these within given time periods. It also required that operations must be carried out 'in accordance with good oilfield practices' and petroleum 'produced in reasonably substantial quantities having regard to the world demand for petroleum and economic exploitation of the petroleum resource'. (Paragraph 25).

A revised, longer draft was prepared in 1954 and this also was submitted to the oil companies for review. It contained the several notable additions and changes noted below. Some of these were undoubtedly made at the behest of the companies which had reviewed the preliminary draft.

The country was divided into four zones for the allocation of concessions, two zones to the north of the 28° parallel and two to the south. In theory, the divisions were based on geographical and geological characteristics but in fact they were partly an attempt to delineate the boundaries between the provinces of the new federal state. As income from any future oil production was intended for provincial treasuries as well as for the federal government, this delineation became a source of friction between the provinces of Tripolitania and Cyrenaica when it turned out that their boundary line split the Sirte Basin with its wealth of oilfields in two.

Secondly, there was a significant reduction in the size of a concession area which a single company was permitted to lease. The new limits in the two northern zones were 30,000 square kilometres and in the southern zones, 80,000 square kilometres. Limits were also placed on the number of concessions a company could hold in each zone.

Thirdly, concession relinquishment terms were left unchanged for the fifth and eighth years, but altered in the tenth year when a company was required to reduce its concession area to one-third of the original size in Zones I and II, and to one-quarter of the original size in Zones III and IV.[33]

Fourthly, the role and membership of the Petroleum Commission was amplified and it was given the power to decide on the granting, assignment, renewal, surrender or revocation of permits and concessions, subject to approval or rejection by the minister of the national economy. It was also empowered to collect fees, rents, royalties and taxes from the oil companies and to appoint a Director of Petroleum Affairs. (Article 2 (3)) Those serving on the commission had to be 'persons of experience in finance, economics, commerce, law or engineering', and members of the Libyan parliament or government were excluded. (Article 2 (2)) Membership was not restricted to Libyans. In fact, A. Hogenhuis, the retired Royal Dutch Shell oilman, was appointed director shortly after the law came into force and held this position until the commission was dissolved in 1963.

One effect of increasing the strength of an independent Petroleum Commission was to dramatically decrease the power of the provincial governors, known as Walis, to influence the oil industry's development. In the earlier draft, each Wali appointed one member of the Petroleum Commission and the three Walis had the power to appoint the Commission's director. (Articles II, III) They also had the final decision on the granting or refusal of concessions. (Article VI) In the revised draft, the Walis had lost all these powers, with the independent Petroleum Commission directed to implement the provisions of the petroleum law 'in the name of each and every Province'. (Article 2 (3))

Fifthly, the thorny question of multiple applications for the same territory was tackled. Applications for concessions were to be awarded on a first-come, first-serve basis, although the Petroleum Commission was given the authority to mediate in cases of conflict, with arbitration required for unresolved disputes. The priority position of early applications made by BP and Shell before the enactment of the law, which the first draft had described as 'deemed to have been made under this Law in the order in which they were made' (Article VII), was abolished. It ruled that all early applications 'shall be deemed to be simultaneous' with those submitted at the appropriate time. (Article 8) But it then complicated the situation greatly by saying that the Petroleum Commission could take into account 'furtherance of the public interest', in awarding concessions. (Article 5 (1) (a)) This opened the door for 'sweeteners', such as increased royalty payments or promises to give money for non-oil related public projects, to influence the commission's decision on applications for the same concession area.

Sixthly, companies were given the right to construct refineries in Libya. (Article 21)

Seventhly, companies were given the right to enter into joint operations for the working of concessions although this required the Petroleum Commission's approval. (Article 14 (4))

Eighthly, companies could claim an annual depletion allowance in the calculation of taxes owed to the government of 25 per cent of gross income up to a maximum of 50 per cent net income. (Article 14 (2)(c))[34]

The final version was completed over the new few months, following meetings between the government and the oil companies. Changes at this time were minor. The only significant alteration was the assignment to the Petroleum Commission of the power to propose any regulations necessary for the implementation of the law. These regulations could cover not only the 'safe and efficient performance of operations' but also the 'conservation of the petroleum resources of Libya'. (Article 24)

What it Achieved

The Libyan Petroleum Law of 1955 was enacted on 21 April, 1955 and the final text, along with two schedules, was published in the *Official Gazette* of 19 June, 1955. Although there were amendments after oil production began, described in Chapter 3, its basic structure has remained unchanged to the present day.

The law contained several unusual features which, in combination, fostered the development of a competitive and prosperous oil industry in Libya throughout the 1960s, a time when there was no shortage of supply elsewhere in the world. Its success in getting oil companies to begin exploration came about, in large measure, from its attractive financial terms. Even a small independent oil company could afford the low entry fee of £500 per concession and an annual rent of £10 to £20 per 100 square kilometres for the first seven years. (Higher rent only began in the eighth year, by which time a company had either found oil in a given acreage or had decided to relinquish it.)

If a company was fortunate enough to discover oil, it had good prospects for appreciable profits. The government's taxation of profits only came into effect when exports were sizeable, and it was based on net profits, after the deduction of rents, fees and royalty payments, as well as expenses and losses, irrespective of where these were incurred. The provisions for allowable expenses were generous. Exploration and prospecting costs could be deducted in the year in which they were incurred or could be capitalized and amortized at an annual rate of 20 per cent and there was an annual depletion allowance of 25 per cent of gross income up to a maximum of 50 per cent net income.

The $12^{1}/_{2}$ per cent royalty charge was calculated not on a posted price but on the 'competitive market price f.o.b. seaboard' minus the cost of handling and transport. Tenure for at least a portion of a concession was guaranteed for 50 to 60 years.

The law also provided companies with assurances which lowered their perception of risk. The creation of an independent, professional Petroleum Commission on which foreigners could serve suggested that issues of economic and commercial concern, rather than politics, would influence decisions regarding the production and exportation of oil. The involvement of oil companies in the drafting of the law boded well for continued good relations between government and companies. In addition, the many foreign and international bodies involved in financial and technical assistance to Libya in the 1950s, with offices in Tripoli and Benghazi, seemed to guarantee political stability.

Two features of the law gave the Libyan government a measure of control over the companies which rose to the bait. By limiting the size of concessions, no one company, or consortium of companies, could lay claim to a large area. And by requiring relinquishment of a considerable part of each concession after a very short time period, companies were deterred from laying claim to acreage and then failing either to explore or to produce oil at a level which suited the government. What the drafters of the law did not foresee was the effect of fostering competition under generous financial terms in conditions where the supply of oil was bountiful and markets were available. This led, inevitably, to a situation where companies were tempted to produce too much too fast, to the detriment of the oilfields.

For more than a decade, lots of oil companies of different complexions took out oil concessions in Libya. They came because of the geology, the proximity of Libya to European markets and the generous terms of the 1955 Petroleum Law. Once they were there and sinking wells, the distribution of small concessions in something like a checkerboard system encouraged exploration efforts by concession holders whenever a company struck oil in a nearby block. The fact that a given percentage of every concession had to be relinquished in a few years and would then be made available as a new concession to others, brought more and more companies in, both independents and majors. By 1968, the government had granted 137 concessions to 39 companies either on their own or in partnership with others, and production was exceeding 2.6 million barrels a day.

Notes

1. The general historical background on Libya is based on the 1960 World Bank Mission report *The Economic Development of Libya*; United Nations, 'Summary of Recent Economic Developments in Africa 1952–53' in *Supplement to World Economic Report* (1955); Petroleum Commission, 'Petroleum Development in Libya 1954 through mid-1961' (1961); Henry Villard, *Libya: The New Arab Kingdom of North Africa* (1956); Abdul A.Q. Kubbah, *Libya: Its Oil Industry and Economic System* (1964); and John Wright, *Libya* (1969).
2. Henry Villard, US Ambassador to Libya from 1951 to 1954, noted that 'if Libya had passed under any form of United Nations trusteeship, it would have been impossible for the territory to play a part in the defense arrangements of the free world'. *Libya: The New Arab Kingdom of North Africa* (1956), pp. 33–4.
3. 'Summary of Recent Economic Developments in Africa 1952–53', United Nations, *Supplement to World Economic Report* (1955); World Bank Mission, *The Economic Development of Libya* (1960).
4. Henrietta M. Larson et al, *History of Standard Oil Company (New Jersey): New Horizons 1927–1950* (1971), p. 733, fn. 40, citing C.L. Locket, 'Middle East and Far East', Coord. Com. Group Meeting, 1947; ibid, A.L. Owens, 'Egypt & Libya'; SONJ, 'Annual Report 1949'.
5. Draft dated 16/7/53 attached file note 2/10/53, BP Archive 9219.
6. A.N. Thomas, 'The Anglo-Iranian Oil Company, Ltd. Minutes of the Annual Geological Conference held at Britannic House, Finsbury Circus, London EC2, 8/9 July, 1952', BP Archive 41659.
7. 'Extension Exploration Permit to D'Arcy Exploration Co. Ltd.', BP Archive 41077.
8. Bernard Mommer, *La Cuestion Petrolera* (1988), p. 13.
9. John M. Blair, *The Control of Oil* (1976), pp. 81–84.
10. Ibid.
11. *Petroleum Press Service*, July 1958, p. 248.
12. *BP Statistical Review of World Energy* (1959).
13. Information on these laws was derived mainly from Benjamin Shwadran, *The Middle East, Oil and the Great Powers* (1955); Stephen H. Longrigg, *Oil in the Middle East* (1961); George Lenczowski, *Oil and State in the Middle East* (1960); Kenneth Redden and Jon Huston, *The Petroleum Law of Turkey* (1956); United Arab Republic, *Petroleum in the United Arab Republic* (1960); and articles in several contemporary journals including *Fortune*, *Economic News* (Tel Aviv), *Middle East Journal*, *Mideast Mirror*, and *Middle East Report*.

14. *Middle East Report*, VI/24, 23 August, 1954. McGhee's experience in the oil industry included positions with Atlantic Refining Company (ARCO) and Continental Oil. He had also served as a partner in DeGolyer, MacNaughton & McGhee and had founded his own oil company.
15. Information on Max Ball from Robert Engler, *The Politics of Oil: A Study of Private Power and Democratic Directions* (1961) and conversations with the American Association of Petroleum Geologists and with Douglas Ball, Max Ball's son.
16. *Mideast Mirror*, V/3, 30 May, 1953, p. 17.
17. Hal Lehrman, 'The Turks like American Capitalists', *Fortune*, June 1954, p. 122; conversation with Douglas Ball.
18. During his government service, Max Ball had been so bitterly opposed to Mexican laws which discriminated against the participation of American firms in Mexico that he had blocked a petroleum development loan requested by the Mexican state oil company, Pemex. Edward W. Chester, *United States Oil Policy and Diplomacy: A Twentieth-Century Overview* (1983), p. 135.
19. *Petroleum Press Service*, April 1954, p. 120; Kenneth Redden and Jon Huston (1956), p. 306.
20. 'Observations Concerning 1954 Drafts of Reconnaissance Permit Form and Petroleum Mining Law and Concerning Comments of Local Representatives Thereon Made at Meeting held in Tripoli June 24, 1954', BP Archive no. 41077.
21. *Middle East Report*, VI/24, 23 August, 1954; *Petroleum Press Service*, June 1954, p. 220; Hal Lehrman (1954), p. 122.
22. United Arab Republic, *Petroleum in United Arab Republic* (1960), pp. 361–8; Benjamin Shwadran, *The Middle East, Oil and the Great Powers* (1985), p. 481.
23. M.D. Schlosberg and L. Kuenstler, 'The Israel Petroleum Law 5712-1952 and Petroleum Regulations 5713-1953', *Economic News* (Tel Aviv), V, 1953, p. 3.
24. In an entry in the 1959 US *Martindale-Hubbell Law Directory*, Batzell is described as author of the Turkish and Guatemalan petroleum laws, but not the Israeli law.
25. I.R. Kosloff, 'Oil Development in Israel: Introduction', *Economic News*, V, 1953, p. iii.
26. Max W. Ball and Douglas Ball, 'Oil Prospects of Israel', *American Association of Petroleum Geologists Bulletin*, 37/1, January 1953, pp. 1–113. Douglas Ball had worked for a number of US oil companies including Phillips Petroleum.
27. Schlosberg and Kuenstler (1953), p. 51.
28. Abdul A.Q. Kubbah, (1964), p. 65.
29. Frank C. Waddams, *The Libyan Oil Industry* (1980), p. 57.
30. Much information on the drafting of the Libyan Petroleum Law was

found in documents in the British Petroleum Archives in Warwick, England.

31. A D'Arcy Exploration Company memo concerning the 1954 draft of the petroleum law noted that 'many other features we believe to be a considerable improvement over the early 1953 draft which we criticized in March of that year'. 'Observations Concerning 1954 Drafts of Reconnaissance Permit Form and Petroleum Mining Law and Concerning Comments of Local Representatives Thereon Made at Meeting Held in Tripoli June 24, 1944,' 5 August, 1954, p. 4, BP Archive 41077.
32. A BP document, 'Comments on a Draft for the Libyan Petroleum Mining Law 1954', BP Archive 41077, p. 9, emphasized the importance of the wording of the section dealing with royalty and market price in order 'to avoid the implication that the Government will by edict "establish" such market price'.
33. Relinquishment requirements were present in earlier concession agreements in Iran and Saudi Arabia. These were of little consequence, however, given the size of the original concession, the small acreage which had to be given up and the number of years concession holders had to decide on which parts to surrender. Those in the Egyptian government's contract with Conorada were much closer to the Libyan requirements.
34. In the United States, a depreciation allowance of 27.5 per cent of gross income or 50 per cent of net income was the general practice. This was intended to compensate companies for the exhaustion of oil reserves. For a discussion of US depletion allowance see Stephen L. McDonald, 'Federal Tax Treatment of Income from Oil and Gas' (1963), pp. 11–16.

3 THE RULES OF THE GAME CHANGE

A host of oil companies rushed to acquire drilling rights in Libya following the passage of the 1955 Petroleum Law. As Appendix 3.1 shows, within the space of five years, 19 companies had secured a total of 84 concessions, mostly in the northern Zones I and II whose entire acreage had been allocated for exploration by the Libyan government. All the major oil companies took some acreage, either on their own or in joint ventures, as did several US independents, including Socal, Phillips, Nelson Bunker Hunt, Amerada, Continental, Ohio (Marathon), Grace, Standard Oil of Indiana (Amoco) and Libyan American (Arco). European companies were also active, including Total, Gelsenberg, Deutsche Erdoel, Wintershall, Elwerath and Compagnia Ricerche Idrocarburi (CORI/ENI). Another 11 concessions were granted in the following 18 months, adding Ausonia Mineria and Elf Aquitaine to the list of non-majors holding acreage. Many of these companies found oil, although not immediately as exploration was delayed by the need to remove several million mines laid during the war. The first few discoveries were not judged to have commercial potential. By 1959, however, the presence of substantial oil deposits in Libya was confirmed.

The 1955 Petroleum Law had been written when there was only a hope – albeit well-founded – that oil would be found in Libya. Not surprisingly, once the existence of abundant reserves was confirmed in the Sirte Basin region, the attitude of the Libyan government changed towards companies seeking and holding concessions. As Morris Adelman astutely noted, a concession agreement drawn up when the results of exploration are unknown inevitably comes under great strain if oil is found. The government involved then wishes it had held out for better terms and, with foreknowledge, the company would willingly have agreed to give these. A rich discovery therefore makes a dissatisfied government, aware that the company's profits are far greater than necessary to keep it producing. It therefore inevitably strives to get a greater share of these profits.[1]

Within a few years of the first oil discoveries, the government announced interpretations of some of the terms of the petroleum law to improve its position. It drafted changes to the law well before the first shipment by Esso of Libyan crude left Brega harbour in September 1961, and throughout the following decade it applied pressure on foreign oil companies to surrender more of their profits and more control of production. At times in the 1960s, the relationship between the government and the companies was distinctly confrontational. When the Revolutionary Command Council seized control of the government in September 1969, it inherited an ongoing policy of increased government control of the country's oil industry.

The government had two substantial advantages in its negotiations with the companies once oil had been found. The first of these was the value of Libyan oil in European markets due to its short haul and its high quality. The second advantage stemmed from the terms of the concession system laid down by the 1955 Petroleum Law. Small concessions available on financially reasonable terms with enforced territory relinquishment in the short term created a situation in which there were companies waiting in the wings to take over acreage which other companies had failed to exploit. The government therefore never felt the need to compromise in order to keep any single company, or group of companies, happy with the conditions under which it held a concession.

The administration of King Idris recognized these advantages, as did its Revolutionary Command Council successor. On the other hand, both administrations realized that they were dependent, in the last analysis, on foreign oil companies to produce and market Libyan oil and gas and that they could not afford to be too inflexible lest there was a general exodus of all companies. The carrot-and-stick policy which such a situation required has continued up to the present day. Commercial considerations tied to production and marketing realities have dictated the development and functioning of the Libyan oil industry. There have been a few instances of decisions taken for political reasons – the nationalization of British Petroleum and of the Amoseas partnership are outstanding examples – but these have been few and far between.

'Extra Benefits for Libya'

The Libyan Petroleum Commission in a report to the Second

Arab Petroleum Congress in 1960, noted that the oil companies which had been granted concessions through 1959 had drilled 122 wells, with 35 of these wells in production. As a result of the enthusiasm of foreign oil companies for concessions, the government explained, it was taking a firmer stand on what it stood to gain from this production. In the report, the Commission explained that, as a result of 'improved prospects and keener competition', it had decided to try 'to improve in moderation the financial arrangements with would-be concession holders and with existing holders'. It said that although there was 'no intention to alter unilaterally contractual rights guaranteed ... this does not at all mean that concession terms should become fossilized'. Changes were necessary because 'a concession deals with a developing situation and it has to develop and grow to meet changing circumstances'.[2]

The government identified the means by which it could attain its goal as Article 5(1)(a) of the 1955 Petroleum Law which instructed the Petroleum Commission to take into consideration 'furtherance of the public interest' when awarding concessions. It was openly encouraging concession applicants to offer 'extra benefits for Libya' in order to succeed in their applications. As the *Petroleum Press Service* noted at the time: 'The eagerness to get areas in Libya is such that though the Petroleum Law does not provide for competitive bidding, would-be explorers, in order to secure bonus-worthy acreage, offer nowadays better terms than the law requires.'[3] Companies bidding for concessions offered a variety of extra benefits to the government. Their 'sweetened' bids included higher royalty payments, higher rent payments, increased drilling obligations, higher minimum investments and the possibility of participation by the Libyan government in production operations at a later date.

The Commission also managed to deny some bidders the benefits of the annual depletion allowance of 15 per cent of gross income up to a maximum of 50 per cent of net income which had been added to the final draft of the 1955 Petroleum Law. The Chairman of the Libyan Petroleum Prices Negotiating Commission later justified this action by explaining that 'such an allowance, while having some merit in conditions in the USA, where is its chief application, is unsuited to the major oil producing and exporting countries of the Middle East'.[4] In 1957,

Amoseas, the partnership of Socal and Texaco, agreed to restrict what it could claim as depletion. In 1959, BP waived its right to a depletion allowance altogether in a concession agreement it was then negotiating as well as retroactively in its past agreements. The Petroleum Commission suggested that 'old concession holders will naturally stand a better chance of increasing their acreage if they follow BP's example'.[5] It left no doubt that a company's failure to drop the depletion allowance in existing concession agreements would influence the commission's decision on that company's application for new acreage.

The Italian company CORI, an affiliate of ENI, made one of the most widely publicized 'sweetened' bids in late 1959. It secured a concession in the face of competition by agreeing to forego a depletion allowance and to pay a royalty of 17 per cent instead of the statutory 12 per cent. If a commercial discovery was made, it would give the Libyan government the option to participate to the extent of 30 per cent in exploration, exploitation, development, marketing and transportation within Libya. In addition, the Libyan government would be represented on the company's board of directors and would have full access to accounts. If the enterprise was extended to cover refining, natural gas extraction or the manufacture of petrochemicals, the government would have participation rights in these operations. (In the event, none of this was realized as CORI failed to make a commercial discovery.)

Although few of the concession agreements signed in the late 1950s and early 1960s were made public, the government let it be known when extra benefits were included. Pan American (Amoco), for instance, conceded limitations on its depletion allowance in a concession agreement and agreed to pay a bonus of $5 million by instalments if and when oil production began. Deutsche Erdoel and Wintershall agreed to spend $1 million on exploration in the first two years of their concession period as well as an extra 2 per cent royalty, provided that the total of such excess royalty did not exceed $1 million. The Elwerath/Deutsche Erdoel/Wintershall consortium agreed to an extra 2 per cent royalty which would not be credited against tax payments. It also agreed that half of its depletion allowance would be spent on exploration. CORI, as noted above, agreed to a 17 per cent royalty and other benefits. Phillips agreed to a 19 per cent royalty and waived its depletion allowance.

The Commission also was able to force acceptance of changes through a clause in the 1955 Petroleum Law which required a company to seek its approval before it could assign part of its concession to another company or buy shares in an existing concession. In order to buy into the Liamco(Arco)/Grace concession, Esso agreed to undertake the construction of a refinery in Libya. Similarly, in order to take a 50 per cent interest in Bunker Hunt's concession, BP undertook to do drilling obligations in excess of the required minimum.

The 1961 Watershed

In August 1960, the government announced that it intended to amend some of the terms of the 1955 Petroleum Law. It noted that in doing so it was following the example of other oil-producing countries, naming Iraq, Saudi Arabia, Kuwait, Qatar and Iran, and adding that the amendments would be 'derived from the most modern and recent provisions applicable in these countries'.[6] It noted that Nadim Pachachi, a former Iraqi minister, was serving as an adviser regarding the revisions and regarding oil policy in general.[7]

The government announced that it would not grant any new concessions until the changes it was making in the law were completed. This pronouncement caused much dismay among companies as considerable acreage in early Sirte Basin concessions had been surrendered in accordance with the terms of the 1955 Petroleum Law, and this acreage, much of it considered to be very promising, would therefore be available in the next bidding round. The companies were concerned, correctly as it turned out, that the government would make acceptance of changes in the law a precondition for the acquisition of new acreage. The Petroleum Commission fostered this concern by noting that the amendments were intended to stimulate competition in areas which had already been relinquished and would 'open the door for the petroleum companies that have acquired concessions in the past to amend their concessions' in order to qualify for bidding when the relinquished areas were offered.[8]

The most significant amendments made in 1961 to the 1955

Petroleum Law were to Article 14 concerning the basis on which royalty was calculated and allowable cost deductions. The amended article stipulated that royalty would be calculated on the posted price rather than on the competitive market price. The definition of posted price appeared in a subsequent Regulation which noted that 'the concession-holder or its affiliates shall from time to time establish and publish its posted price for Libyan crude oil'. Oil companies interpreted this phrase as giving the Petroleum Commission's agreement to their unilateral posting of prices; the Libyan government disputed this interpretation. It intended to increase its royalty income by having the final say over the posted price level. As will be seen in Chapter 6, ambiguity on this point became the cause of serious conflict between the government and the companies.

Article 14 was also altered to ensure that the government received, as originally intended by the 1955 Petroleum Law, a 50 per cent share of oil company profits through a combination of fees, rents, royalties and taxes. In order to achieve this level, it felt it was necessary to curtail the costs allowed as deductions from gross income to arrive at a declaration of taxable income.[9] The right to a depletion allowance, which had already been curtailed in many concession contracts, was formally abolished. Annual amortization of pre-production intangible exploration expenses was reduced from 20 per cent to 5 per cent and annual depreciation of pre-production expenditure of physical assets was reduced from 20 per cent to 10 per cent.

There were also amendments at this time to Articles 7 and 8 of the law intended to legitimize the inclusion of 'extra benefits'. It was stated that concessions would be granted by a process of competitive bidding and that 'the additional conditions and extra economic and financial benefits offered by applicants have now become the basic consideration in selecting the successful bidder'.[10] (At a later date, the government published a list of 17 'Elements of Preference' which could influence its decision on concession awards. All of these involved financial returns to the government over and above what was required by law.)

The Royal Decree which brought these 1961 amendments into effect stipulated that no new concessions would be awarded to applicants holding prior concessions unless they agreed to amend agreements for these in accordance with the changes.

Marketing Expenses Row

The amended Article 14 of the 1955 Petroleum Law defined the basis on which taxes would be levelled as the posted price of Libyan crude petroleum less marketing expenses multiplied by the number of tons of crude oil exported. But it did not spell out exactly what marketing expenses covered, and this turned out to be a serious omission. The Libyan government apparently assumed that claimed marketing expenses would not exceed 2 per cent of posted price, which was the current practice among the major oil companies with Middle East production.

Esso, which had Middle East acreage as well as Libyan acreage, went along with this assumption. Marathon, Continental and Amerada, the independent oil companies which made up the Oasis partnership, did not. They argued that 2 per cent reflected the insignificant marketing expenses of the major oil companies which distributed their crude oil production through existing integrated marketing affiliates. The Oasis partners did not have built-in integrated markets and insisted that the only market for their Libyan production was non-affiliated refiners in Europe. In the highly competitive market conditions of the early 1960s, these refiners were demanding substantial discounts.

The Oasis partners managed to persuade the government that they would not be able to achieve the large-scale production of Libyan oil which the government wanted unless they offered discounts. The government obliged by issuing Regulation 6 in November 1961 which defined marketing expenses as 'the sum of expenses which were fairly, properly and necessarily attributable to the sale, organization and transportation of petroleum for export from Libya and other services related thereto'.[11] These expenses included 'the sum total of rebates if any from the posted price which the concession holder is obliged to grant for the purpose of meeting competition, in order to sell Libyan crude petroleum to affiliated or non-affiliated customers'.[12]

This regulation effectively sanctioned the Oasis practice of heavily discounting the posted price to buyers in Europe, thereby greatly reducing their taxes. In effect, Marathon, Continental and Amerada managed this way to pay taxes not on the posted price but on the realized price of their crude production. Their average declared net price for tax purposes between 1962 and

1964 was $1.58/barrel and at one point fell to $1.30/barrel. In the meantime, Esso was paying taxes on its posted price for Libyan oil of $2.12/barrel minus marketing expenses of 2 per cent. Regulation 6 sanctioned a two-tier basis for the calculation of royalty which made the major oil companies then producing oil in Libya – Esso and the Amoseas partners of Texaco and Socal – very angry. In the end, the independent companies paid for their favoured position. When the Qaddafi government, a decade later, put pressure on them to pay higher taxes, the majors refused to help them.

OPEC Membership

In its explanation regarding the 1961 amendments to the 1955 Petroleum Law, the Petroleum Commission noted, in an oblique reference to OPEC, that 'This is what has actually taken place in other countries such as Venezuela and the Arab countries where the agreements were amended to the advantage of these countries, as a result of negotiations.' [13] Libya had joined OPEC in 1962, two years after the organization had been founded in order to stop the major oil companies from reducing their posted prices of Middle East oil and thus the basis of taxable income for governments of producing countries. Libya shared OPEC's stand on the need to maintain government income from the taxation of oil production. However, it did not want to provoke a confrontation on the issue of posted price levels with the oil companies holding concessions in Libya at this time, fearing that doing so might slow the momentum of the industry's development.

In its first years as a member of OPEC, Libya was essentially a 'free-rider', glad to profit from the organization's gains in negotiations with oil companies but unwilling to endorse OPEC policies if they would alter the government's relationship with the companies that had started production in Libya. In principle, it endorsed OPEC's support of increased refinery capacity in producing countries, the establishment of national oil companies, an integrated approach to oil industry operations and co-ordination of the conservation, production and exploitation of oil resources. Its upstream policies in the early and mid-1960s were generally in accord with the official line of

OPEC. But by 1968, it was openly critical of some OPEC views and, in the early 1970s, it had important differences regarding price levels and the nationalization of foreign oil companies.

The question of the accounting treatment of royalty payments was one OPEC stand in the early 1960s which Libya openly endorsed, very much to its own advantage. The practice in the Middle East at the time was for royalty payments to be fully credited against income tax liability, thus reducing the amount of taxes owed. In 1962, OPEC passed a resolution recommending its members to insist that royalty payments should be 'expensed', that is to say, treated as a cost item. The Iranian and Saudi Arabian governments began negotiations with the major oil companies producing in their countries to this effect, but made little progress. In August 1963, talks began between the Secretary General of OPEC and a committee appointed by the Iranian Consortium. The oil companies offered to accept royalty expensing if they were allowed a 12.5 per cent discount off posted prices for tax purposes, an offer which effectively nullified any government tax gain as a result of royalty expensing. Eventually, after prolonged negotiations, the companies' demand for a discount was lowered to 8.5 per cent, to be gradually reduced over three years to 6.5 per cent, when it would be reconsidered. This offer was accepted by all the Middle East OPEC producers except Iraq.

The major oil companies were obliged to submit this offer to Libya, as a member of OPEC. Esso and the Amoseas partners did so in November 1964. They used the opportunity to bring the issue of the marketing discounts claimed by the independent oil companies to the fore, citing the disparity in declared taxable income which was the result of this practice. Esso and Amoseas insisted that although the independent companies had not been involved in the OPEC Middle East royalty expensing negotiations, they must also agree to expense royalty payments. If not, Esso and Amoseas threatened to introduce high marketing expenses in their taxation calculations, on the grounds that they were being forced to discount in order to retain their share of the European market which was being undermined by the independents' price-cutting.

The government, too, saw the royalty expensing issue as the

means, which it had long sought, to rectify a situation where at least two-fifths of Libya's exports were being sold at highly discounted rates, resulting in a considerable loss of its taxation income. It issued a memorandum in October 1965 to all the oil companies with concessions in Libya, noting that:

> The government is now of the opinion, on information received, that the independents have had sufficient time to find market outlets for most of their oil. Further, it is known that they have also now become financially associated with refineries in Europe. Bearing these factors in mind the government has come to the conclusion that it simply cannot afford to continue allowing the present system of rebates ... by certain companies; not only is it having an adverse effect on the Libyan economy, it is destroying the European market for oil products; it is killing the goose that lays the golden egg.[14]

The memorandum went on to say that the government had decided to negotiate with all oil companies with a view to introducing 'the more equitable OPEC formula, that is the expensing of royalties'.

These negotiations, if they occurred at all, were hardly prolonged. In November 1965, the government again amended the 1955 Petroleum Law by Royal Decree. The major changes at this time were that:

i. Royalties would be treated as an expense of production and not as payment in advance of income tax.
ii. Income for tax purposes would be based on posted prices less the discount agreed between OPEC and the major oil companies which was 7.5 per cent for 1965, 6.5 per cent for 1966, and no higher than 6.5 per cent subsequently.
iii. Marketing expenses allowed as a deduction from income could not exceed 0.5 US cents per barrel.

In an accompanying memorandum, the government indicated that it was making these changes 'in order to apply the OPEC settlement in Libya; it aims at rectifying the unacceptable position of discounts and reductions and at increasing the payments which the companies undertake to make to the government'.[15] It cited Esso's stipulation that the major oil companies would agree to expensing of royalties only if it was 'applied to all companies without exception'. To soften the blow, the government offered a relief regarding certain types of prior taxes due to the government but unpaid.

By the middle of January 1966, all the oil companies working in Libya had agreed to amend their concession agreements. They expected that the government would then finally announce the award of new acreage; none had been granted since 1961.

New Entrants

By the time the holders of existing concessions were reconciled to the expensing of royalties, the government was sitting on more than 100 bids from 50 companies for new concessions which included hitherto unassigned areas, as well as almost 200,000 square miles of relinquished acreage.

Awards of new concessions were announced in February 1966. The 'extra benefits' offered by successful bidders included bonuses to be paid when production reached a certain level, the promise of government participation in production operations, higher tax payments and undertakings to fund and oversee industrial projects, such as a refinery or a petrochemical plant. Forty-one new concessions were allotted; the last in 1968.

The government's intention to strengthen connections with West European companies was evident in these awards. West German companies did exceptionally well. Gelsenberg, in partnership with Mobil, acquired three concessions and the partnership of Elwerath and Wintershall got four. In addition, Union Rheinsche won four concessions and Scholven Chemie, three. West European state-owned companies also did well. The Italian state-owned Agip obtained two concessions. Two French companies, Elf Aquitaine, a mixed state enterprise, and Auxerap, a subsidiary of state-owned Erap, combined with Hispanoil, a Spanish state-owned enterprise which later became Repsol, and Murphy Oil Corporation, a US independent, to form a joint venture in two concessions. Subsequently, Elf Aquitaine obtained a concession on its own. The government favoured European state-owned oil companies in the hope that these would assure Libya secure, long-term markets, given the preference of government owners for stability as well as their interest in profits which often meant a source of foreign exchange. This policy has continued to the present day and the involvement of state-owned companies in Libya undoubtedly has tempered the political attitudes of their governments towards the Libyan government.

By comparison, several US companies which had expected concessions were disappointed. Esso and Oasis both failed to get new awards; Esso, however, was later able to acquire a 50 per cent interest in existing concessions. Among the majors, Mobil, Shell and the Amoseas partners succeeded in some of their applications whereas Gulf and BP had not applied for new acreage. Several new US independent companies appeared on the scene, including Occidental, which later discovered oil and developed production on a large scale in acreage relinquished by Oasis. (In order to obtain this acreage, for which 20 companies were said to have bid, Occidental committed itself to spend 5 per cent of pre-tax profits on an agricultural scheme in the Kufrah Oasis, a place of symbolic importance to King Idris, and to build a $30 million ammonia fertilizer plant, whose cost would be shared with the government. It is also said that Occidental wrapped its bid in the colours of the Libyan flag.)[16] As Appendix 3.2 shows, several small unknown US companies also received concessions despite doubts of their abilities to fulfil the legal requirements for exploration activities and investment; several later had their awards revoked for this reason.

Stronger Federal Government

A significant reorganization of the structure and functions of the government followed the amending of the Petroleum Law. In April 1963, the central government felt sufficiently secure to embark on measures that would concentrate power in its hands. It abolished the federal structure of the country, dropped the word 'United' from the title 'Kingdom of Libya', and curtailed the power and responsibilities of the three provincial governments.

The central government's role in the oil sector was enhanced shortly thereafter by a law which abolished the Petroleum

Commission and transferred all its rights and responsibilities to a Ministry of Petroleum Affairs organized into technical, legal, accounting, research, economics and administrative departments. The law also established a High Council of Petroleum Affairs, consisting of various ministers, the Governor of the Bank of Libya and three outside members with relevant professional experience. The council's powers did not include any executive or other decision-making authority without reference to the Council of Ministers.

The administration and implementation of the amended Petroleum Law and the development of the Libyan oil industry was now clearly in the hands of the central government with the help of advisers. One effect of this change was to tie more closely the government's plans for the expenditure of funds, particularly its development projects, to taxation of oil production.

One of the first decisions of the newly strengthened Ministry of Petroleum Affairs was to establish a Libyan national oil company and it sent an observer to a meeting in November 1966 of the state oil companies of Iran, Iraq, Saudi Arabia, Kuwait, Indonesia and Venezuela to look into this matter. At least some of the governments whose companies were represented at this meeting intended to assume complete control of their oil industries by nationalizing foreign operations. The Ministry of Petroleum Affairs did not want to go this far in its participation in the development of the Libyan oil industry. It was concerned by reports of discussions at the conference as to whether state oil companies of producing nations could secure market share in a period of weak demand and soft prices without discounting and giving rebates, practices which could further lower prices and, therefore, government income from taxation.

The government announced the establishment of the Libyan General Petroleum Company (Lipetco) in April 1968, noting that the company had been set up 'with a view to entering into partnerships with others on the basis of new terms which will be more beneficial to Libya'.[17] The government intended to keep full control of the new company. By law, the Council of Ministers had the final authority to decide on any acreage allocated to Lipetco and to approve all its participation agreements as well as its annual budgets.[18] Lipetco would assume the government's share in any joint oil exploration and development ventures and its right to participate in all existing and future concessions.

End of Concession Agreements

In July 1968, the Idris government announced that it would not make any more concession agreements and that all new acreage would be awarded within the framework of joint ventures with Lipetco. In this way, it explained, the state would acquire equity interest in producing concessions. Lipetco concluded its first joint venture with the French state-owned companies, Erap and Elf Aquitaine in 1968. Under the terms of this agreement, the government was to have a 25 per cent interest in the venture with its share to be increased up to 50 per cent when production reached approximately 560,000 b/d. Erap and Elf Aquitaine would bear all exploration costs whether or not oil was found, but development costs would be shared in proportion to the partners' interests. The companies would be liable for income tax based on posted prices. The Libyan government hoped that it would be able to persuade more European state-owned oil companies into similar joint ventures which could amount, in some cases, to government-to-government agreements.

In the months that followed, several companies made proposals for joint ventures with Lipetco, and in May and June 1969 four agreements were announced, with Shell, ENI (Agip), Ashland Oil and Refining and Chappaqua Oil. As usual, 'extra benefits' for Libya were involved. Shell agreed to arrange financing and construction of a 25,000 b/d refinery near Tripoli with a 600 b/d lubricating oil blending plant and ENI agreed to build a natural gas pipeline to supply Benghazi and a number of service stations. Ashland agreed to build a carbon black plant and Chappaqua paid a bonus and promised to construct pipelines. (The Qaddafi government later declared the Chappaqua joint venture null and void on the grounds that it was in breach of the Petroleum Law on several counts, reimbursing Chappaqua for its bonus payment but not for its exploration expenses.)

Increased Pressure on Companies

Libya's formation of a national oil company and its success in forcing concession holders to base their tax liability on posted prices with a minimum of discounts, gave the government a sense of increased power over the oil industry. The impressive growth of oil production and exports during the latter half of the 1960s encouraged it to press for advantages in negotiations with oil companies.

As the decade drew to a close, the government was determined to try to increase its revenues from oil production, convinced that its per-barrel take lagged behind that of other oil-producing countries. Early in 1969, it drew up two regulations which would give it considerably more control over the industry. The first of these, Regulation 8, concerned oil and gas production levels. One of the consequences of the generous financial terms for companies laid down in the 1955 Petroleum Law, combined with the enthusiasm in European markets for Libyan oil, was that the companies were doing very well. In late 1965, for instance, two Esso executives were reported as saying that their company's production in Libya was its 'most profitable' of anywhere in the world.[19] The companies knew that they were on a good thing and could not be sure how long it would last. They were therefore producing as fast and as much as they could from their concessions.

Libya feared, probably justifiably, that overproduction of oilfields by concession holders was occurring, to the detriment of the long-term recovery of oil reserves. The government was therefore very pleased at a suggestion made at a 1968 OPEC conference that members should adopt and implement a 'Pro Forma Regulation for the Conservation of Petroleum Resources' calling for close supervision of exploration and production operations to ensure that these were based on conservation practices required in the USA, Canada and Venezuela. Regulation 8 required oil companies to submit detailed, advanced information on proposed drilling operations to the government for approval. In addition, it gave the government the option of taking, free of charge at the wellhead, any natural gas not being used for 'economically justified' industrial or commercial purposes, a measure designed to encourage exploitation of natural gas by the companies. This regulation, which became law without consultation with the oil companies, provided the government with a legal basis for specifying reduced production levels per field on the grounds of conservation. Libya did not seek to reduce production in order to influence world market prices as it realized that its volume of oil exports was small in relation to total exports. It knew that monopolistic attempts regarding the market on its part easily could be defeated by expanded production in Iran, Saudi Arabia and Kuwait, for example. The issue was genuinely the application of best techniques and practices to optimize long-term recovery rates. On the other hand, production restraints also later proved to be a very effective method of bringing pressure on oil companies to agree to changes desired by the Qaddafi government.

Regulation 9 was an equally forceful statement of government authority over the oil industry. It transferred the authority to set posted prices, which the government consistently objected to as being too low, from the companies to the government. Other provisions of this regulation included the redefinition of what constituted direct taxes, allowable operating expenses and overheads and an explanation of a new procedure for fixing pipeline tariffs. A draft was sent to the oil companies in January 1969 with a request for their comments. A revised version was submitted to them in June for comments or acceptance by the beginning of September. In August, a government spokesman said that posted prices should be raised by 10 cents to reflect the freight and quality differentials of Libyan crudes compared to other Middle East crudes. He added that unless this was done, the oil companies might be ordered to indemnify the government for losses resulting from under-priced exports since 1961; the companies insisted that the government had given them official clearance of prior tax payments up to 1965.

Final negotiations between the government and the oil companies on this regulation were scheduled to start in September. The government warned that Regulation 9 would become law in its existing form if the oil companies failed to reach agreement by the end of the year. The negotiations were postponed following the military coup of 1 September, which resulted in the overthrow of the existing government and the installation of the Revolutionary Command Council headed by Colonel Qaddafi.

The new government did not immediately announce any drastic changes in Libyan oil policy; official statements emphasized that it would honour existing agreements and pledged that there would be no nationalization. On the other hand, the government insisted that it intended to safeguard the interests of the Libyan people by 'a more effective control' over oil operations. In October 1969, *Petroleum Press Service* noted that:

> The old regime had already proved itself an increasingly tough negotiator and was pressing hard for a revision of the petroleum regulations at the time of its overthrow ... On paper at least, the new men are hardly likely to be less so, and renewed pressure for higher posted prices, along with expansion of Lipetco's activities, can be expected. (p. 384)

Participation Agreements

Negotiations over oil prices, which are discussed in Chapter 6, occupied the whole attention of the new Libyan government and OPEC in 1970 and 1971. By mid-1971 this issue had quieted down sufficiently that efforts could be directed to the question of participation. In August 1971, the Libyan government succeeded in obtaining the agreement of Shell to revise an earlier concession contract as a participation agreement. In the meantime OPEC, which several years earlier had issued a formal declaration calling for its members to acquire 'a reasonable level of participation in concessions' called a meeting in September to discuss participation. The outcome of this meeting was OPEC Resolution 139 instructing all member countries to negotiate with their oil companies regarding participation.

In 1972, OPEC issued its General Agreement on Participation. Despite the fact that Nadim Pachachi, the former adviser to the Libyan government on oil policy, was Secretary General of OPEC at the time, Libya did not agree with many of the terms of this agreement which called for government participation to start at 25 per cent and to rise to 51 per cent in 1982.

Although its early joint venture with Erap and Elf Aquitaine had been along these lines, Libya now wanted immediate 51 per cent participation. The government was in close liaison with Algerian authorities at this time and the Algerian oil minister was known to have visited Tripoli seeking support for his country's dispute with France.[20] Libya was undoubtedly influenced by the Algerian decision to nationalize 51 per cent of its French-owned concessions in February 1971.

In addition, Libya disagreed with the OPEC General Agreement in that it wanted compensation at net book value, not updated book value. It also wanted higher buy-back prices (the price at which companies were obliged to buy back a portion of the government's new equity oil) than those specified in the agreement.

In September 1972, it persuaded ENI (Agip) to convert its arrangements for the acreage containing its Bu Attifel field, with an estimated production level of 250,000 b/d, from a concession award to a 50 per cent participation agreement. The government, in return, paid 50 per cent of approved expenditure already incurred by Agip in this acreage. It was only after the agreement was signed that the government approved the start of commercial production and the export of crude oil from Bu Attifel.

In January 1973, the government launched its drive for participation with the independent oil companies then working in Libya. It sought agreements similar to that concluded with Agip, involving an immediate 50/51 per cent government interest, compensation based on net book value of assets, and a buy-back price halfway between tax-paid cost and posting price, less commission. Negotiations continued until August, when the three independent companies in Oasis (Continental, Marathon and Amerada Hess) and Occidental agreed to 51 per cent Libyan participation with compensation based on original net book value and high buy-back prices. The tough terms which the Libyan government was able to secure reflected the strong position of Libyan crudes in European markets, the fact that these independent companies lacked oil reserves outside of Libya and soaring US oil imports.

The government then turned back to the major oil companies which were still resisting its demands. On 1 September, 1973, the fourth anniversary of the revolution, it issued a general decree nationalizing 51 per cent of the assets and business of the major and independent oil companies, and their consortium partners, which had not already consented to participation. Gelsenberg and Grace accepted promptly, but the others delayed, with the outbreak of the Arab–Israeli war in October deflecting the government's attention from this issue for several months. In March 1974, Mobil and Esso signified their willingness to accept 51 per cent Libyan participation; Shell refused and its share in the Oasis group was taken over by the government.

There had been, however, some losses. In September 1970, Gulf surrendered three concessions in western Libya where it had failed to develop commercial production, and left the country. In December 1971, the government nationalized BP's half share of Concession 65, with its Sarir field. The nationalization was a political move at variance with commercial interests; it was intended as retaliation for the failure of the British government, which then held 48.6 per cent of BP, to prevent Iran's seizure of three islands in the Persian Gulf. The government subsequently demanded that Bunker Hunt, which held the other half share of Concession 65, market BP's share of Sarir crude and accept participation.[21] The company's refusal, partly because it lacked any marketing outlets of its own in Europe, resulted first in the government blocking its exports and eventually, in June 1973, to the nationalization of the Bunker Hunt holdings.

In February 1974, the government reacted to the US invitation to the heads of government of the major oil consuming countries to a Washington Energy Conference to discuss united action against what it considered to be 'unprecedented price rises by OPEC which threaten the monetary and economic stability of the world'.[22] In retaliation, Libya seized the operations of the Amoseas partnership of Texaco and Socal (Chevron). These covered 14 blocks totalling 45,000 square kilometres with the partnership's main output coming from the Nafoora field. It also seized Arco's Libyan American share of Esso Sirte, involving three blocks of 9,000 square kilometres whose only production was from the Raguba field, operated by Esso. Although these actions, like the nationalization of BP and Bunker

Hunt, were designed to protest against US Middle East policies, they were commercially sound. The government was able to take over these companies and to continue their operations without any marked effect on the health of the industry as a whole.

Although the government won the battle of participation, it was at a cost of losing, to a great extent, the trust and confidence of foreign oil companies. This was clearly evident in their caution regarding expansion. The total number of rigs active in exploration and appraisal declined from 55 in 1969 to only eight in 1973, and the number of wells completed fell from 327 to 72 over the same period. Within a few years, the government realized that it had to come up with new arrangements to encourage companies already in Libya, and new entrants, to undertake further exploration and development.

In Retrospect

The 1955 Petroleum Law succeeded admirably in attracting many different oil companies to look for oil in Libya. Once they began to make important discoveries in which they had vested interests, the government was able to take a tougher line on its share of the winnings. It attempted to do so within a legal framework by first issuing interpretations of the law which were to its benefit and then by amending certain provisions of the law to give the government more powers. The main structure of the 1955 law, however, remained intact.

The government's motives in changing the law were financial in nature. It wanted to increase its oil revenues but leave the business of producing and selling oil in the hands of foreign companies. Even in its most radical periods, it only sought 51 per cent participation, never complete nationalization. The few takeovers which did occur were motivated by events outside the oil industry and did not set a precedent.

Perhaps the most important and lasting success of the 1955 Petroleum Law was its ability to attract companies other than the majors. This allowed the Libyan oil industry to continue almost without a hitch when the US companies were ordered to leave in 1986, much to the chagrin of the US government.

It also had its faults. Allowing the inclusion of 'extra benefits for Libya' in the awarding of concessions in the 1960s had the potential of undermining the legal structure which the drafters of the 1955 law had so carefully devised to regulate the oil industry. It may also have played a role in bringing about the downfall of the monarchy. That several totally unknown oil companies were awarded acreage in 1966 suggests that the system was being corrupted. The Qaddafi government's later exploration and production-sharing agreements, described in Chapter 4, were intended partly to prevent 'sweeteners' from influencing acreage award decisions. They were mostly successful on this count, but failed to stimulate exploration which remained at a low ebb for more than 20 years.

Notes

1. M.A. Adelman, *The World Petroleum Market* (1972), p. 42.
2. Petroleum Commission, *Petroleum Development in Libya 1954 through mid-1960* (1960), p. 7.
3. *Petroleum Press Service*, May 1960, p. 164.
4. *Middle East Economic Survey*, IX/ 9, 31 December, 1965, p. 3.
5. Petroleum Commission (1960), p. 8.
6. Petroleum Commission, *Petroleum Development in Libya 1954 through mid 1961* (1961), p. 8.
7. Pachachi joined the Directorate of Petroleum in Iraq in 1934 and for some 25 years was either directly or indirectly connected with oil developments in the country. He held ministerial positions for several years and represented Iraq on many rounds of negotiations with IPC. For his involvement in Libya, see Abdul A.Q. Kubbah (1964), p. 78.
8. Ibid.
9. Petroleum Commission (1961), pp. 12–16.
10. Petroleum Commission, *Petroleum Development in Libya 1954 through 1962* (1962), p. 6.
11. Kingdom of Libya, Ministry of Petroleum Affairs, *Libyan Oil 1954–1967* (1968), p. 42.
12. Ibid.
13. Petroleum Commission (1961), p. 16.
14. The memorandum was apparently compiled by two public relations firms, Sydney-Barton of London and James F. Fox of New York. See 'Libya's Case for New Oil Law', *Middle East Economic Survey*, Supplement, IX/ 9, 31 December, 1965, p. 2.
15. 'Libyan Petroleum Law Amendment: Memorandum Explaining the Draft Law Embodied in a Royal Decree, Amending Certain

Provisions of the Petroleum Law,' *International Legal Materials Current Documents*, V/ 3, 3 May, 1966, p. 442.

16. John Wright (1981), p. 226.
17. *Petroleum Press Service*, May 1968, p. 183.
18. See 'Full Text of Law Establishing Libya's National Oil Company', *Middle East Economic Survey Supplement*, XI/ 26, 26 April, 1968, pp. 1–9.
19. Adelman (1972), p. 67, citing *Platt's Oilgram News Service*, 8 November, 1965, p. 4.
20. 'The International Oil "Debacle" since 1971', *Petroleum Intelligence Weekly Supplement*, 22 April, 1974, p. 23.
21. Concession 65 had originally been granted to Bunker Hunt which had signed a 60-year partnership agreement with BP prior to the discovery of the Sarir field. Following the BP nationalization, Hunt had no legal agreement with the Libyan Arabian Gulf Exploration Company formed to operate the Sarir field and its takeover was not unexpected. See *Petroleum Press Service*, July 1973, p. 245.
22. Mr. William Simon, head of the US Federal Energy Office, had said that the USA was likely to seek a reduction in oil prices at this meeting to mitigate the damage to the balance of payments of all nations heavily dependent on oil imports. *Petroleum Economist*, February 1974, p. 50.

APPENDIX 3.1

Concessions Awarded Through 1961

Concession Number	*Holder*	*Date of Grant*
1	Esso 100%	11/55
2	Nelson Bunker Hunt 100%	11/55
3- 8	Esso Standard 100%	12/55
9-15	Mobil 65%, Gelsenberg 35%	12/55
16-17	Liamco (Arco) 25.5%, Grace 24%, Esso 50%	12/55
18-19	Liamco (Arco) 51%, Grace 49%	12/55
20	Liamco (Arco) 25.5%, Grace 24%, Esso 50%	12/55
21-22	Liamco (Arco) 100%	12/55
23-24	Total 100%	12/55
25-33	Oasis - Amerada 33 %, Continental 33 %, Marathon 33 %*	12/55
34-37	BP 100%	01/56
38-41	Shell 100%	01/56
42-47	Texaco 50%, Socal 50%	12/55
48	Esso 100%	05/56
49	Total 100%	05/56
50	Mobil 65%, Gelsenberg 35%	05/56
51	Texaco 50%, Socal 50%	05/56
52	Shell 100%	12/56
53-54	Continental 100%	12/56
55-56	Oasis 100%*	12/56
57	Mobil 65%, Gelsenberg 35%	01/57
58	Esso 100%	11/56
59-60	Oasis 100%*	01/57
61	Total 100%	03/57
62	Mobil 65%, Gelsenberg 35%	02/57
63-64	BP 100%	07/57
65	Nelson Bunker Hunt 50%, BP 50%	12/57
66-68	Gulf 100%	04/57
69-70	Shell 100%	12/57
71	Oasis*	12/57
72	Mobil 65%, Gelsenberg 35%	12/57
73	Texaco 50%, Socal 50%	11/57
74-76	Pan American (Amoco) 100%	03/58
77	Deutsche Erdoel 50%, Wintershall 50%	08/58
78	Elwerath 33 %, Deutsche Erdoel 33 %, Wintershall 33 %	06/59
79	Gulf 100%	08/59
80-81	BP 100%	09/59
82	CORI 100%	11/59

Concession Number	*Holder*	*Date of Grant*
83	Texaco 50%, Socal 50%	12/59
84	Pan American (Amoco) 100%	12/59
85	Ausonia 60%, Deutsche Erdoel 20%, Elf Aquitaine 20%	03/60
86-89	Liamco (Arco) 100%	06/60
90-92	Phillips 100%	04/61
93-95	Pan American (Arco) 100%	04/61

* In 1966, Amerada assigned 50% interest in its share to Shell.

APPENDIX 3.2

Concessions Awarded 1966–68

*Concession Number**	*Holder*	*Date of Grant*
96-99	Wintershall 50%, Elwerath 50%	03/66
100-101	Agip 100%	03/66
102-103	Occidental 100%	03/66
104-105	Elf Aquitaine 28%, Hispanoil 42%, Auxerap 14%, Murphy 16%	04/66
106-109	Union Rheinische	04/66
110-111	Phillips 100%	04/66
112	Mercury 100%	04/66
113	Lion 100%	04/66
114	Shell 100%	04/66
115-117	Scholven Chemie 100%	04/66
118	American Mining 100%	04/66
119-120	Clark 100%	04/66
121-123	Circle Oil 100%	04/66
124-126	Mobil 65%, Gelsenberg 35%	05/66
127	Libyan Desert 100%	05/66
128-130	Libyan Texas 100%	05/66
131-133	Texaco 50%, Socal 50%	07/66
134-135	Bosco 100%	07/66
136	Arco 50%, Phillips 50%	12/66
137	Elf Aquitaine 100%	04/68

* Concession numbering began again in 1966; concession 102 or concession 103, taken by Occidental, for instance, contained acreage surrendered from concessions 25-33 taken by Oasis.

Source: Kingdom of Libya, Ministry of Petroleum Affairs, *Libyan Oil 1954-1967* (1968), Table 16.

4 PRODUCTION-SHARING CONTRACTS

In July 1973, the Libyan government issued a decree announcing that all future arrangements with foreign oil companies would be in the form of production-sharing agreements. Oil industry contracts involving production sharing had been offered first by the Indonesian government in the mid-1960s and their use spread to several other OPEC producers. Under this type of legal arrangement, the government, usually through its national oil company, retained exclusive title to the acreage involved but had the authority to conclude agreements for petroleum operations within this acreage with oil companies which were then considered as contractors. The companies provided the necessary funds for exploration, carried the risk and generally acted as operators of the acreage in fact, if not in name. The classic production-sharing contract allowed the company a share of production for the recovery of its development costs; in Indonesia, where the concept of cost recovery originated, the first 40 per cent of production went to the company to cover costs. In other countries, the share varied, generally between 20 per cent and 40 per cent. The production which remained after costs had been deducted was then divided between the company and the government in accordance with a proportion set out in the agreement, with the company generally subject to income tax on its share. As with cost recovery arrangements, the production split varied in different countries.

Libyan production-sharing agreements up until 1988 differed from the majority of such agreements concluded elsewhere in that they did not set any part of the production aside for cost recovery. Instead, the oil company received a fixed percentage of production, free of income tax and other charges. This percentage was different for offshore and onshore acreage and was a subject for negotiation in every contract signed. Once a discovery was made, the Libyan government, through NOC, would advance part of the development costs. The company was then obliged to repay the funds advanced when production began, usually at a rate of 5 per cent per annum. In some cases, outstanding amounts were partly or wholly interest bearing, with the rate of interest fixed at one point below the international market rate as defined by the Central Bank of Libya.[1]

EPSA I

The first Libyan exploration and production-sharing agreements, known as EPSA I, were made in 1974. In February of that year, the Revolutionary Command Council, in its first award of exploration acreage since coming to power, signed an EPSA I with Occidental for 19 blocks, mostly in the Sirte Basin. The agreement was to run for 35 years unless no commercial discoveries were made, in which case it would expire in five years. The agreement stated that the Libyan National Oil Company (NOC) which had replaced Lipetco 'shall be entitled to take and receive 81 per cent of the crude oil produced hereunder and Occidental 19 per cent'.[2] Occidental was to bear all exploration costs, expenses and obligations with development expenses to be borne initially by both parties in proportion to their respective entitlement to production. Occidental, however, was obliged to repay NOC's share of development costs at a rate of 5 per cent a year, without interest, over a 20-year period when production reached a stipulated but undisclosed level. Its share of income realized from production was exempt from all fees, rents, royalties, income taxes and surtaxes. Four years later, Occidental had made six oil discoveries in this acreage.

By 1978, Libya had signed similar EPSA I contracts with Exxon, Mobil, Total, Elf Aquitaine, Braspetro and Agip. The standard production split was 85:15 for onshore blocks and 81:19 for offshore acreage. Exxon and Mobil signed EPSA I for new acreage shortly after they accepted 51 per cent participation in 1974. Agip acquired both onshore and offshore blocks; the latter turned out to contain the Bouri oilfield and a large gas field. A consortium of Elf Aquitaine (75 per cent), Wintershall (10 per cent) and NOC (15 per cent) was given two offshore blocks close to the Agip acreage. It found some oil in these blocks but did not develop this as Elf Aquitaine effectively withdrew from activities in Libya throughout the 1980s – partly as a result of political strains in Libyan–French relations due to Libya's involvement in Chad and partly because it had holdings in Tunisia and did not want to become involved in the Libyan/

Tunisian offshore boundary dispute.[3] Total acquired offshore and onshore acreage which did not prove profitable. Braspetro signed an EPSA I for blocks in the Sirte and Murzuq basins; it found oil in the Murzuq acreage but in a small field lacking gas and water pressure and it relinquished the acreage in the early 1980s.[4]

The government varied its involvement with the activities of the various companies involved in exploration and production in Libya throughout the 1970s in accordance with its interests in their acreage. It never negotiated production-sharing agreements, for instance, for a small onshore operation of a consortium of Elf Aquitaine, Hispanoil, Auxerap and Murphy, which eventually was relinquished, or with Wintershall.

EPSA II

By 1979, the government had become seriously concerned with reserve depletion and the need for new discoveries of oil. Its first move to encourage exploration was to demand the relinquishment of tracts then held by foreign companies which were not being exhaustively explored. Elf Aquitaine, Agip, Mobil, Total and the Oasis consortium complied with this demand to some extent, and the area under licence declined from about 783,000 square kilometres at the end of 1978 to 645,000 at the end of 1979.[5] The government then proceeded to offer much of this acreage under EPSA II contracts. Despite the fact that the terms of EPSA II were even less favourable to foreign companies than those of EPSA I, the government reported considerable interest. In 1980, it announced that it was negotiating EPSA II with 13 various foreign groups but did not elaborate. Occidental, Elf Aquitaine, the Oasis consortium and Shell took on more acreage and the US company Coastal States was awarded four blocks, including one offshore block. The West German firm Deminex, a joint venture of Veba Oel, RWE-DEA für Mineraloel und Chemie and Wintershall signed an EPSA II contract for the exploration of five onshore and one offshore block. It relinquished four of these, retaining one Sirte Basin block where it discovered a small oilfield in 1982, and one Ghadames Basin block.[6]

Another aspect of Libyan oil policy at this time was to try to involve east European state oil companies in upstream activities in Libya, a decision which was not popular with NOC. The latter had misgivings regarding the financial ability of these companies to undertake development of oil which they discovered, doubts which turned out to be well founded. The Romanian state oil company, Rompetrol, which signed an EPSA II for a block in the Murzuq Basin in south-west Libya, made important discoveries which it was unable to develop due to lack of funds. The Bulgarian state oil company, Geocom, signed an EPSA II for blocks in the Murzuq and Ghadames Basins and also was unable to proceed with their development.

US Companies Depart

The government's problems with insufficient exploration were not solved by the EPSA II contracts, which failed to attract many new companies to chance their luck in Libya, despite the availability of what many in the industry viewed as promising acreage. The situation became more critical when the US government ordered all US companies and US citizens to leave Libya in 1986. The companies affected by this decree were Conoco, Marathon and Amerada Hess, which had holdings of 16.3 per cent, 17.3 per cent and 8.2 per cent respectively in Oasis; Occidental with a 36.75 per cent stake in its participation agreement with NOC; and Grace with a 12 per cent holding in the Raguba field.

The United States had started placing restrictions on trade with Libya as early as 1978 as it suspected the Libyan government of involvement in the support of terrorist activities. In 1982, it banned the import into the USA of Libyan crude oil. This did not seriously affect Libyan production as the government had little difficulty finding west and east European markets to absorb the exports formerly intended for the USA. Moreover, the ban exempted crude oil sent to the Amerada Hess refinery in the Caribbean. Similarly, the 1985 US ban on imports of Libyan products did not have a significant effect on the Libyan oil industry.

The US government's recall of all US oil companies in June 1986 had more implications for the Libyan oil industry. In the short term, it affected exploration more than production, as there were many Libyans who had gained production experience through working with the departed US companies. Moreover, the Libyan government considered that the US companies continued to retain their claim to these concessions which they would be able to exercise at some future date. This meant that the companies had a vested interest in the wellbeing of their Libyan oilfields and were willing to offer advice on production problems when this was requested.

The ousted US oil companies and Libyan government representatives held sporadic meetings to try to settle the legal status of their holdings in the years that followed. Initially, the government gave them 'suspended rights' to their acreage for three years. When this period of grace approached its end in mid-1989, the US government suggested that it might soften its position and allow the companies to continue their operations in Libya by transferring the entities to foreign-named organizations. The Libyan government refused to accept this type of arrangement, but continued to be in contract throughout the early 1990s, despite occasional rumours, denied by the government, that Libya was looking for non-US buyers for these assets.

Most of the major companies that held exploration and production acreage in the early years and later departed were American companies. Those that went of their own accord for financial and other reasons included Gulf and Phillips which left in 1970, Amoco which left in 1975, Esso which surrendered all its Libyan operations including its LNG plant in 1981 and Mobil which left in 1982. The holdings of Nelson Bunker Hunt were nationalized in 1973 and those of the Amoseas partnership of Texaco and Socal in 1974. The US companies that still remained were forced to leave by the US government in 1986. Of the non-US majors, BP was nationalized in 1971 and Shell finally left in 1991.[7]

The Libyan National Oil Company established four subsidiaries to manage the acreage left behind by these companies. The Arabian Gulf Corporation (Agoco) was formed in 1980 as a merger of the earlier Arabian Gulf Exploration Company (Ageco), which took over BP/Hunt concessions, and the Umm Jawabi Oil Company, which took over the Amoseas concessions. The Sirte Oil Company was established in 1981 to manage the Esso operations including those connected with natural gas. After the 1986 departures, the Waha Oil Company was set up to take over the operations of the Oasis holdings and the Zueitina Oil Company, a partnership of NOC and OMV, to handle the acreage of the former Occidental holdings.

EPSA III

Although the government stopped issuing public data on reserves and production, there were few signs of significant reserve additions in the 1970s and 1980s. It would appear that most drilling resulted in the discovery of small accumulations that were not worthwhile developing commercially, especially after the collapse of oil prices in the mid-1980s. The exceptions were the successes made by NOC which found the Messlah field, by Agip which found the offshore Bouri oilfield and NC-41 gas field and which enlarged its Bu Attifel reserves, and by the east European state oil companies which made discoveries in the Murzuq Basin.

By the end of the 1980s, government revenue had fallen to the point where the discovery of a larger reserve base and increased oil production became imperative. Low oil prices, low production, the effect of sanctions, and the costs of constructing the Great Man-Made River and the iron and steel works at Misurata were the main causes of financial problems. To remedy this situation, the government announced, early in 1988, that it was offering new, improved exploration and production-sharing agreements for onshore and offshore acreage to foreign oil companies. The terms of EPSA III were designed to be more attractive than those of the earlier EPSAs. For the first time, cost recovery, found in most production-sharing agreements elsewhere, was included, with an agreed percentage (to be negotiated separately for each contract) allocated to the contracting oil company until its cumulative value equalled cumulative expenditures.[8] The agreements were intended to yield an acceptable return by international standards and to ensure a rapid pay-out for the foreign firms. They attracted a number of companies. By the end of 1995, there were some 25 foreign oil companies involved in exploration in Libya. Most of these were newcomers and many were simply investors of funds in existing consortia. Several, however, had been active in Libya in both exploration and production for a number of years. Foreign oil companies had flocked to Libya following the passage of the 1955 Petroleum Law and a foreign presence in upstream Libyan operations was still very much in evidence 40 years later. Like the monarchy it replaced, the Revolutionary Command Council sought outside investment of funds and expertise, although companies were not always willing to respond.

Long Libyan Experience

Agip is by far the most important foreign oil company in Libya; its production in 1995 was running at 260,000 b/d of crude oil.[9] Agip first became involved in Libya in the late 1930s, when the Italian government instructed it to undertake a survey for petroleum reserves. After the end of the Second World War an Agip affiliate, Compagnia Ricerche Idrocarburi (CORI), was awarded a concession in which it was never successful; Agip later took over this acreage which yielded discoveries of three fields in the 1990s. In 1966, Agip acquired two blocks in the last round of concession offers; one of which it relinquished. In 1968, it discovered the Bu Attifel field in the other block and subsequently, the North Rimal field; by the mid-1990s its Bu Attifel complex was producing 190,000 b/d.

In 1977, Agip discovered the offshore Bouri field and significant gas reserves in nearby offshore block NC-41. It began producing from Bouri in the late 1980s and by the mid-1990s was planning to move on to stage two of the development of this field; this was designed to increase production capacity from 70,000 or 90,000 b/d to 150,000 b/d. It was also discussing development of the gas field in offshore block NC-41 in terms of an eventual export of 8–10 bcm/year of natural gas by pipeline to Sicily. Realization of these plans was not immediately forthcoming and the Libyan government became concerned that Agip was not proceeding quickly with the development of all its Libyan oil and gas potential because of the extent of its holdings elsewhere in North Africa.

Agip has been involved in assorted legal agreements with Libya. In 1973, it agreed to convert its existing concession agreement for the Bu Attifel block into a participation contract and later signed an EPSA I for new acreage. A subsequent

EPSA II contract, with a production split of 81:19, covered Agip's offshore acreage containing Bouri and the NC-41 gas field. In 1994, Agip signed an EPSA III contract for all its holdings under previous, less favourable EPSA as well as for five new blocks, four in the Ghadames Basin in western Libya and one in the Sirte Basin. The production split for the offshore acreage reportedly was revised upwards from 81:19 to 70:30.

OMV, the Austrian national oil company, is a second company with a long history of working in Libya, having first become involved in 1975, when Elf Aquitaine took it as a 15 per cent partner, along with Wintershall (10 per cent), to help with the costs of exploration and development of an offshore block where an oil discovery had been made. In 1984, it renegotiated the terms of its participation in this consortium and established a subsidiary, OMV of Libya Limited. The following year it purchased 15 per cent of the equity interest of Occidental. When Occidental left in 1986, OMV and NOC formed a joint venture, the Zueitina Oil Company, for the oversight of this acreage which is theoretically still under Occidental's aegis. NOC, the operator, held 87.5 per cent of Zueitina Oil, and OMV 12.5 per cent. In May 1994, Abu Dhabi's International Petroleum Investment Company (IPIC) acquired a 19.6 per cent stake in Zueitina Oil which had made an oil discovery in block NC-105 in its holdings in 1992.

In 1989, OMV signed an EPSA III, with Braspetro and Husky Oil as partners, covering new acreage.[10] It was designated operator in a block in the western Ghadames Basin, where it had a 40 per cent interest, as well as in three blocks in the Sirte Basin, where it had a 51 per cent interest. In the fourth Sirte Basin block, where Braspetro was designated operator, it has only a 30 per cent share. By the early 1990s, OMV had interests in a total of 13 blocks in Libya, with an exploration area of 33,500 square kilometres. In 1992, it declared a well in one of its Sirte Basin blocks commercial, listing reserves of 22.8 million barrels. In late 1994, it took a 30 per cent share in the Repsol/Total consortium that had recently signed a contract to develop the three oilfields in block NC-115 in the Murzuq Basin which had been discovered by Rompetrol. In 1995, OMV's average production in Libya, apart from its equity shares, was 10,000 b/d.[11] This figure was expected to increase considerably with the coming on stream of NC-115, scheduled for 1997.

Veba, the German oil and petrochemical company, is another Libyan veteran which became involved in EPSA III contracts. Veba began operations in Libya in 1978 when it took over the minority share of Gelsenberg in concessions which it had held with Mobil since 1958. These have estimated oil-in-place reserves of 4 to 5 billion barrels with a 30 per cent recovery factor and include the Amal field. When Mobil withdrew from Libya in 1982, Veba became the operator of Amal. In 1988, Veba increased its share in the former Mobil/Gelsenberg concessions from 17.15 per cent to 49 per cent through an EPSA III agreement. In addition to Amal, the other producing fields in which Veba has an interest include Ora, Hofra and the Ghani/Zainat/Zala fields. In 1994, it made an oil discovery near the Amal field. Its production from its holdings increased in the early 1990s, largely as a result of upgrading facilities in the Amal field and the start of a water flooding project in the Ghani/Zainat/Zala fields. At the end of 1995, production was around 90,000 b/d.[12]

Wintershall is another German company with a long history of working in Libya. It has an unusual legal status in that it has never signed an EPSA and continues to work under its original concession agreements. Wintershall took a half share with Deutsche Erdoel in a concession in the south-western Sirte Basin in 1958; no oil was discovered in this acreage which was relinquished in 1971. In 1959 it took a share, with Deutsche Erdoel and Elwerath, in another concession in the same area. In the next round of concessions which began in 1966, Wintershall, in partnership with Elwerath, was awarded four blocks in the central Sirte Basin; oil was discovered in this acreage in 1970. As noted above, Wintershall, along with OMV, entered into a consortium headed by Elf Aquitaine in 1975.

In the early 1990s, Wintershall was the sole concessionaire and operator of the central Sirte Basin blocks which it had been awarded with Elwerath; one of these is considered to contain proven reserves of 144 million barrels, principally in the As-Sara and Jakhari fields. The Hamid and Ruama fields, in other blocks, are considered to contain 94 million barrels proven reserves. In 1995, Wintershall brought on-line a 3000 b/d well at its Nakhla field, raising its output to nearly 60,000 b/d.[13]

Total, formerly Compagnie Française des Pétroles (CFP), has been intermittently active in Libya over the years. In 1955 and 1956, in the first tranche of concession awards, it received four concessions which it relinquished as unprofitable in 1970. Four years later, Total signed an EPSA I for several blocks in the Murzuq Basin in south-west Libya and offshore. Once again, it was not successful in finding commercial oil in this acreage and the company effectively withdrew from exploration in Libya throughout the 1980s.

In the early 1990s, Total re-entered Libya in force. In 1993, it signed an EPSA III to develop the Mabruk field in the centre of the Sirte Basin which had been discovered by Exxon in 1959 but left undeveloped due to the highly fractured nature of the reservoir and the low gas–oil ratio of the field. Development costs for Mabruk have been estimated at $1 billion and Total is rumoured to have arranged a 70:30 split of production after costs are covered.[14] The field has oil-in-place estimated at around 1.3 billion barrels of 35–38°API crude with a 0.26 per cent sulphur content. Total is planning a phased development of this acreage; pilot phase work began in February 1995 and by the end of the year, production had increased from an initial 2500 b/d to 10,000 b/d. In August 1995, the first export shipment was made from the Es Sider terminal.[15] Production is scheduled eventually to reach 50,000 b/d as the result of an extensive water injection programme.

In 1994, Total joined the consortium headed by the Spanish oil company Repsol to develop the NC-115 block in the Murzuq Basin discovered by Rompetrol and estimated to have recoverable reserves of some 1 billion barrels. Repsol is another Libyan oldtimer which became very active in Libya in the 1990s. In 1966, Hispanica de Petroleos SA, the precursor of Repsol, entered into a partnership with Elf Aquitaine in two Sirte Basin concessions. These were eventually relinquished as non-productive and Hispanica de Petroleos took no part in any of the early EPSA offerings.

In October 1994, Repsol assumed a significant involvement in the Libyan oil industry by signing an EPSA III for the development of the A, B and H oil accumulations in the Murzuq block NC-115, covering 4,275 square kilometres. Rompetrol had drilled 57 wells before relinquishing this block, estimated to contain approximately 1 billion barrels of recoverable reserves of high quality 43.1°API crude oil with a content of less than 0.2 per cent sulphur and no metals. Repsol is the operator, with a 20 per cent share. Total and OMV both have 15 per cent shares and the rest is held by Arabian Gulf Oil Company, a subsidiary of NOC. Despite its reluctance to offer more than a 85:15 production split in known fields, the government gave Repsol and its partners close to the 75:25 arrangement which they sought, with the consortium's share free of royalty and taxes. The EPSA III provides for a 50:50 split between the consortium and NOC of development costs, estimated at around $1 to $1.5 billion, with NOC paying for its share after the start of production. The consortium will pay 25 per cent of the operating costs and there is no ceiling on cost oil for cost recovery purposes.[16]

Production was scheduled to start in NC-115 in late 1996 at an initial rate of 50,000 b/d, increasing to 85,000 b/d in the following year, with a target of 100,000 b/d by the year 2000. A substantial water injection programme will be needed as well as a 400-kilometre, 30-inch pipeline from the field to Hammadah al-Hamra where it will link up with an existing 380-kilometre, 18-inch pipeline to the Zawiyah refinery and export facilities on the Mediterranean which was built by NOC in the 1980s. In August 1995, Repsol awarded a contract of $155 million to a Cypriot company, Joannou and Paraskevides, for the supply and construction of this pipeline. When production reaches its target, additional pipeline facilities will be needed.

Newcomers

The government signed several EPSA III contracts after 1988 with companies that had not worked previously in Libya and subsequently allowed the signatories to farm out parts of their holdings to other investors. These companies, like those which have remained in Libya for many years, generally have widespread, worldwide exploration and production interests, thereby spreading their risks. Most have given high priority in the selection of acreage for exploration to the proximity of existing pipelines and export facilities. They are unwilling to consider the expenditure of large sums on infrastructure construction in order to transport any oil they might find to coastal ports for export or distribution internally.

One of the first new arrivals was the International Petroleum Corporation of Vancouver, British Columbia, which established a wholly owned subsidiary, International Petroleum Limited, to work in Libya.[17] IPC signed an EPSA III in April 1989 covering three blocks, NC-154, NC-155 and NC-156, whose acreage totalled 932 square kilometres. These were located in the Sirte Basin close to the Amal, Intisar and Augila oilfields. The government agreed production sharing at a 70:30 split and the company was allowed cost recovery. In August 1991, IPC farmed out shares in these blocks to two British companies, Lasmo (London and Scottish Marine Oil) and Hardy Oil and Gas. IPC remained the operator with a 45 per cent interest; Lasmo's stake was 40 per cent and Hardy's 15 per cent. Seismic studies and drilling, however, did not prove promising and this acreage was apparently relinquished in 1995.[18] Earlier that year, Hardy had announced its intention to withdraw from this operation in conjunction with its restructuring plans aimed at concentrating its activities outside of Africa and continental Europe.

In May 1992, IPC signed a second EPSA III for three other blocks with a total acreage of 21,884 square kilometres, NC-176 and NC-177 in the Sirte Basin and NC-178 in the barely explored Cyrenaican platform. Preliminary exploration identified several Palaeocene reef prospects in NC-176, located near the Intisar field, but an exploratory well drilled in 1995 was abandoned as a result of poor results. IPC was designated operator of these blocks. Originally it held a 75 per cent share, with Hardy Oil & Gas having 15 per cent and Sands Petroleum of Sweden, 10 per cent. By 1995, IPC's share had been reduced to 40 per cent, Sands increased to 40 per cent and a newcomer, Seven Seas Petroleum held 5 per cent. Hardy announced its intention to withdraw from this arrangement in that year.

INA-Naftaplin, a Croatian firm, signed an EPSA III in July 1989 for five separate blocks in the Sirte Basin with a total area of 1279 square kilometres.[19] The terms of this EPSA were very generous, with production sharing reportedly agreed at a 65:35

split, cost recovery allowed for the company and a requirement for the expenditure by INA-Naftaplin of only $28 million over five years. This was a striking example of the government's policy of bringing east European companies into Libya despite their limited financial ability to undertake development of any oil discoveries. INA-Naftaplin signed a second EPSA III agreement in December 1989 covering one block of 2250 square kilometres located offshore, north-west of Tripoli. The production-sharing split was once again 65:35, with cost recovery for the company and an investment requirement of $16.8 million over five years. Croatian companies involved in Libyan construction projects outside of the oil sector were in considerable financial difficulty following the outbreak of the Yugoslav civil war; presumably INA-Naftaplin experienced the same problems.

As noted earlier, when OMV signed an EPSA III in 1989 for one block in the Ghadames Basin and four blocks in the Sirte Basin, it took in two partners for this venture. One partner was Braspetro and the other, Husky Oil International of Calgary, Canada. Husky took a 30 per cent share in the Ghadames Basin block and in the largest Sirte Basin block. Husky, with a 49 per cent share, was the only partner with OMV, the operator, in the other three smaller Sirte Basin blocks.

In 1990, Petrofina made a major investment in exploration in Libya. Its EPSA III contract signed in February involved three blocks located in the central and southern Sirte Basin, with a total acreage of 15,736 square kilometres.[20] Petrofina's commitments for these blocks were considerable, involving 8500 square kilometres of seismic, 12 exploratory wells and an expenditure obligation of $100 million by 1996. This was the largest single financial commitment to date by a foreign company in the EPSA III round.

In 1992, Petrofina, operating in Libya as Fina Exploration Libya, farmed out shares in these blocks, thus reducing its participation and risk to 50 per cent. The original farm-out was to three Canadian companies, Westcoast Petroleum of Calgary which received 7.5 percent, Chieftain International of Edmonton which received 7.5 per cent and Pan Canadian of Calgary which received 15 per cent. Westcoast was subsequently taken over by Sirocco Energy, a Hongkong consortium headed by Numack Energy. Later that year, three Korean companies bought into the venture, Yukong with a 10 per cent interest, Hyundai with 5 per cent interest and Lucky Goldstar with 5 per cent. By early 1995, the only report of an oil discovery was in block NC-171.

A British company involved in exploration projects in the early 1990s was Lasmo, through its subsidiary, Lasmo Grand Maghreb Ltd. As noted above, Lasmo took a 40 per cent interest in IPC's first three Sirte Basin blocks which were relinquished in 1995. It also signed an EPSA III on its own behalf for two blocks offshore (NC-173 blocks 1 and 2) in the Gulf of Sirte covering 23,632 square kilometres and one block (NC-174) in the south-western Murzuq Basin covering 11,310 square kilometres. Its partner in this venture, with a 50 per cent interest, was a consortium led by the state-owned South Korean company Petroleum Development Corporation (Pedco) with a 12.5 per cent interest. The other members of this consortium included Hyundai with a 12.5 per cent share, Daewoo with a 12.5 per cent share, Majuko with a 7.5 per cent share and Daesung with a 5 per cent share. Lasmo was designated operator for which the total exploration investment requirement is $60 million.

Saga, the Norwegian private oil company, took a 25 per cent share of Total's arrangements in the Mabruk field, mentioned above, in late 1994. Saga was scheduled to invest $36 million in the first three-year phase of development of Mabruk and in the mid-1990s was said to be discussing new acreage with the Libyan government.

North African Petroleum Limited, a private UK-registered company owned mainly by Arab business interests, signed an EPSA III in 1991 for two blocks in the western Sirte Basin near the Mabruk field with a 70:30 production split. This is a small venture, and the company is committed to invest only $10 million in exploration activities.

7th November Zone

The Libyan government made an unusual award in September 1988 when it gave the 3000 square kilometre block entitled 7th November for exploration and development to a joint Libyan–Tunisian Exploration and Exploitation Company. For many years there had been a dispute between Libya and Tunisia regarding their offshore boundary, especially after the discoveries by Agip of the offshore Libyan Bouri field and NC-41 gas field and by Elf Aquitaine of the offshore Tunisian Miskar gas field. At one stage, hostilities reached the point where gunboats were sent out to prevent exploration. Eventually, the two countries agreed to submit the matter to the International Court of Justice which established a boundary line in 1982. Tunisia, however, did not accept this ruling and further ICJ hearings had to be held before a boundary acceptable to both sides was agreed in 1985.[21]

After the final ICJ adjudication, Libya and Tunisia agreed to undertake joint exploration of the continental shelf area traversing the offshore boundary. The 7th November zone (the name commemorates the date of the assumption of the Tunisian presidency by Zein al-Abidine Ben Ali) covers 1600 square kilometres on the Libyan side and 1400 square kilometres on the Tunisian side. It is believed to contain three oil- and gas-bearing structures, two small ones on the Tunisian side and a major one, the Omar structure, on the Libyan side. Recoverable reserves have been reported as high as 4 billion barrels of oil and 114 billion to 280 billion cubic metres of gas.[22] Seismic surveys and analyses of Omar have revealed geological similarities to Bouri field.

According to the agreement, the costs of exploration and development are to be shared by both countries, although the intention from the beginning apparently was to offer the zone on an EPSA III basis for development following preliminary seismic exploration which began in late 1991. Although foreign companies were asked for bids for involvement in November 1992, no agreements had been concluded by the end of 1995, apparently because the Tunisians and Libyans differed in their views regarding the type and amount of foreign investment needed. It was rumoured that Libya, which considered itself a significantly important oil province, sought a major exploration deal whereas Tunisia was willing to consider smaller contracts involving the amassing of data.

EPSA IV?

In the early 1990s there were rumours that the Libyan government was going to offer EPSA IV contracts which would include valuable acreage then held by NOC, possibly some of the holdings which nominally continued in the hands of US companies. Nothing had been offered by the end of 1995 amidst speculation that the rumours had been started simply to generate interest, and that the government had no intention of announcing an EPSA IV round. It was said to be very reluctant to release acreage held in trust for US companies because it believed that a solution would be found in the near future whereby these would be able to take up their operations. The government undoubtedly would like these US companies, with their expertise and financial assets, to return and it would certainly not want to become embroiled in legal proceedings by having given their holdings to others.

NOC was also said to oppose giving up some of its acreage to foreign companies despite the fact that it lacked funds for exploration itself. It apparently feared that it would be seen as incompetent if an incoming company made a significant discovery in acreage where NOC had failed to find anything.

On the other hand, the government did not rule out the possibility of an EPSA IV round in the 1990s. In the meantime, it continued to offer EPSA III contracts and companies continued to be interested.

The Effect of Past Success

With hindsight, it is possible to see a gradual evolution of legal arrangements for foreign companies involved in Libyan oil operations from the original concession agreements granted in the 1960s through the participation deals of the 1970s and the subsequent production-sharing contracts. Some companies have experienced the full measure of these changes, which were no doubt seen by most at the time as hostile. Others have been offered only the more recent kinds of contractual agreements. What is apparent from a review of all these legal frameworks is the fact that the Libyan government, despite the differences between the philosophies behind the monarchy and the Revolutionary Command Council, has never lost sight of its need for outside assistance to develop its oil and gas industry to its full capacity. Moreover, when necessary, it has compromised in order to get help.

Given this basic willingness on the part of the government to allow foreign companies to participate, what has kept the companies themselves interested, as Chapter 5 indicates, is the country's oil and gas reserve potential. The early successes, however, in a sense have had an unfortunate effect on subsequent activities. The discovery of exceptional giant fields in the Sirte Basin area led to the construction of extensive export facilities for these, including long pipelines across the desert and coastal terminals with enlarged harbours. The consequence has been to limit most subsequent exploration to areas close to existing infrastructure. In the 1980s, only East European companies looked further afield onshore and, ironically, they found in the end that they lacked the funds to develop and export their discoveries.

Exploration in the Sirte Basin area since 1980 has been disappointing. It may be necessary for the government to offer more attractive production-sharing terms in order to persuade companies to look farther out.

Notes

1. For a discussion of production-sharing contracts, including those of Libya, see Gordon H. Barrows, *Worldwide Concession Contracts and Petroleum Legislation* (1983), pp 9–13, 16–17.
2. For text of major clauses of the agreement, see American Society of International Law, *International Legal Materials, Current Documents*, XIV/8, May 1975.
3. In the mid-1990s, Elf Aquitaine was reported to be considering further exploration in its acreage.
4. In 1989, Braspetro joined a consortium with OMV and Husky Oil which had interests in blocks in the Sirte and Ghadames basins; it was designated operator in one of these blocks.
5. *Petroleum Economist*, January 1981, p. 30.
6. It was inactive in both of these blocks for many years, but was reported in 1993 as recommencing exploration.
7. Shell was largely unsuccessful in Libya. It failed to find commercial oil in its original concession acreage which it later relinquished. In 1965, it acquired a half interest in Amerada's share of Oasis but lost this through government action in 1974. It signed an EPSA II agreement, later amended to an EPSA III in 1980 for a block in the Ghadames Basin. In 1990, it relinquished this block and took an interest in an adjacent block held by Soc. Energia Montedison under an EPSA III. This block was relinquished in 1991.

8. Article 12.1A, 'Model Exploration and Production Sharing Agreement of 1990,' *Barrows Petroleum Legislation, Supplement 109,* July 1992, Appendix II.
9. This was some 20,000 b/d lower than 1994 production as a result of 'natural decline' in the Bouri field. Agip expressed confidence at the time that infill drilling and a subsea development well would reverse or at least stabilize Bouri's output. See *Petroleum Intelligence Weekly*, 1 May, 1995, p. 5; 28 August, 1995, p. 5.
10. OMV, 'Annual Report 1984', p. 12; 'Annual Report 1985', p. 20; 'Annual Report 1989', p. 35; 'Annual Report 1990', p. 18.
11. Salomon Brothers, 'OMV: Buy: Accelerating Pace of Recovery', *European Equity Research: European Oils* (1995), p. 26.
12. Veba, 'Annual Report 1988', p. 46; 'Annual Report 1991', p. 60; 'Annual Report 1992', p. 56; Veba Oel, 'Annual Report 1993', p. 38; *Petroleum Intelligence Weekly,* 29 August, 1995, p. 5.
13. *Petroleum Intelligence Weekly*, 28 May, 1995, p. 5.
14. *Petroleum Intelligence Weekly,* 1 August, 1994, p. 1.
15. *SAGA News*, no. 3, 1995, p. 9; *Middle East Economic Survey*, 38/47, 21 August, 1995, p. A12.
16. For the terms of this agreement, see *Petroleum Intelligence Weekly*, 33/31, 1 August, 1994, p. 1 and *Middle East Economic Survey*, 30/4, 24 October, 1994, p. A6; 38/46, 11 August, 1995, p. A14.
17. IPC is owned by the Lundin Group.
18. See Yorkton Securities, *International Petroleum Corporation*, 1 September, 1995, p. 10.
19. Blocks SS with 257 square kilometres, OO with 265 square kilometres, HH with 299 square kilometres, KK with 211 square kilometres, and FI with 247 square kilometres.
20. NC 170 contains 7081 square kilometres, NC 171 contains 6876 square kilometres and block NC 172 contains 1779 square kilometres.
21. For details of the ICJ boundary see *Middle East Economic Survey*, 27/47, 3 September, 1984, p. A3. The ICJ also ruled in June 1985 on the boundary between Libya and Malta.
22. *Petroleum Argus*, 2 November, 1992 p. 7.

5 UPSTREAM: RESERVES AND PRODUCTION

Libya was indeed fortunate in its geology, having four (some say three) sedimentary basins holding promise of oil deposits. The most important of these, from the point of view of known oil reserves, is the Sirte Basin, or Sirtica as it is often called. A barren desert, the Sirte Basin historically separated the inhabitants of the western Libyan province of Tripolitania from those in the eastern province of Cyrenaica. It covers 375,000 square kilometres, stretching 500 kilometres south from the Gulf of Sirte into the Sahara desert and 700 kilometres from east to west. It was formed during transgressions and regressions of the sea across the relatively flat North Africa Shelf during prehistoric times and has several distinct uplifted platforms and deep troughs running in a north-west direction beneath its surface of flat rough sand, gravel pavements, salt flats, small mesas and relatively few extensive sand dunes.[1] An estimated 1.3 billion square kilometres of late Mesozoic and Tertiary marine sediments have accumulated in these troughs and on these platforms.

As the map at the beginning of the book shows, almost all of the Libyan crude oil fields which have been discovered to date are located in the Sirte Basin, mostly in large reservoirs which, in 1995, held 72 per cent of Libya's reserves of original oil in place. There were 16 giant Sirte fields under production at this time, each with original reserves of more than 500 million barrels of oil. These are mostly carbonates with shale caprock, but the giant fields in the south-east of the basin – Sarir, Messlah and Bu Attifel – are sandstone of Cretaceous or Cambrian age, overlain by thick marine Cretaceous shales; oil is produced directly from the sandstones.[2] Most early Sirte discoveries were at depths of between 3000 and 5700 feet and the yields per well were good or, in some cases, exceptional. The oil found was very light, ranging from 35°API to 46°API, with a low sulphur but, in some cases, a high wax content.

Aside from the giants, most of the Sirte Basin's other producing fields have original listed reserves of between 100 and 300 million barrels. Very few smaller fields have been discovered, which is unusual in a sedimentary basin of this size.

This partly reflects a lack of incentive to find and develop small fields. The foreign oil companies in Libya in the 1950s and 1960s had their hands full developing the large fields which they discovered. They had little need for more reserves in smaller fields, nor probably the capacity to deal with such. After the Qaddafi government came to power, lower profits and insecurity of tenure reduced investment by these companies in exploration. In the 1980s, the Libyan government claimed that OPEC production quotas did not leave room to bring new reserves into production. A more significant reason for reduced exploration activity was the fact that foreign companies, and even the Libyan National Oil Company itself, tended to restrict their activities to areas convenient to existing pipelines and export infrastructure in order to keep development costs at a minimum.

Aside from the Sirte Basin, the other sedimentary basins of Palaeozoic origins in Libya are the Murzuq and Ghadames Basins which lie to the south and south-west, and the Kufrah Basin which lies to the south-east. These received only cursory exploration by oil companies in the heady early days of the development of the Libyan oil industry but, with the exception of Kufrah, became more important in exploration and development later on.

Murzuq is a large, ladle-shaped structure 600 kilometres long and 1000 kilometres wide situated in the former province of Fezzan, specified as Zone IV in the 1955 Petroleum Law's territorial division of the country for the purpose of allotting concessions. It is separated from its northerly extension, the Ghadames Basin in Zone I, by the Al Qarqaf platform; some consider the Murzuq and Ghadames Basins to be a single sedimentary basin. They have similar geological features to the nearby Illizi (former Polignac) Basin of Algeria, including the same sandstone deposits from the late Jurassic and early Cretaceous periods.[3] Murzuq's remoteness from the coast and its inhospitable terrain, with sand dunes that sometimes exceed 300 metres in height, discouraged systematic exploration in the early days. There was more activity in the Ghadames Basin where several oil companies sunk wells and found oil but considered their finds unprofitable to develop. Later, the Libyan National Oil Company (NOC) discovered several small oilfields and proceeded with their development. In the 1980s, east

European oil companies made significant discoveries of oil reserves in the Murzuq Basin, including a giant field with an estimated 2 billion barrels of original oil in place.

The third large Libyan sedimentary basin, Kufrah, in south-east Libya, Zone III, is about 500 kilometres long and 900 kilometres wide and has structural elements of varying ages and trends. Although it contains Palaeozoic strata which are partly marine, it appears to lack suitable hydrocarbon source rocks and there have been no significant oil discoveries here to date.[4] A more hopeful onshore location for oil reserves is the Cyrenaican platform in the north-east corner of Libya, Zone II. This has a limestone surface but also contains heavy sedimentation from the Palaeozoic, Mesozoic and Tertiary periods.[5] Very little exploration has occurred to date on this platform.

An important area for oil and gas reserves is the Tripolitanian Shelf, known locally as the Pelagian Shelf, which lies off the Libyan and Tunisian coasts west of Tripoli. The Bouri oilfield, with 2 billion barrels of original oil in place, and the Block NC-41 gas field both lie in this shelf, as do other smaller fields. To the east, there have been some minor offshore discoveries in the Gulf of Sirte and off the Cyrenaica platform which, as yet, have not been proved commercial.

It seems highly likely that Libya contains undiscovered oil and gas reserves within and outside the Sirte Basin. Many believe that undiscovered reserves may lie predominantly in stratigraphic rather than anticline reservoirs. Finding such reservoirs, however, is difficult and expensive and will undoubtedly require foreign capital and know-how. The indications are that the Libyan government realizes this.

Reserves

Libya's official proved reserves, listed in Table 5.1, increased dramatically from 3 billion barrels in 1961 to 35 billion barrels in 1969.[6] In 1971, the Qaddafi government lowered the official reserve estimate from 35 to 25 billion barrels, arguing that oil companies had inflated their field reserve figures in order to justify overproduction. Except for a rise in 1972 which was subsequently rescinded, the government kept this 25 billion barrel figure, with minor adjustments, until 1986, when it reached 22.8 billion barrels. It remained at this level for the next four years despite subsequent production and the discovery of new reserves after this date.

Table 5.1: Libyan Proved Recoverable Crude Oil Reserves. Billion Barrels

Year	*Reserves*	*Year*	*Reserves*
1961	3.0	1978	24.3
1962	4.5	1979	23.5
1963	7.0	1980	23.0
1964	9.0	1981	22.6
1965	10.0	1982	21.5
1966	20.0	1983	1.3
1967	29.2	1984	21.1
1968	30.0	1985	21.3
1969	35.0	1986	22.8
1970	29.2	1987	22.8
1971	25.0	1988	22.8
1972	30.4	1989	22.8
1973	25.5	1990	22.8
1974	26.6	1991	22.8
1975	26.1	1992	22.8
1976	25.5	1993	22.8
1977	25.0	1994	22.8

Source: *OPEC Annual Statistical Bulletin, BP Statistical Review of World Energy.*

In May 1990, the Libyan oil secretary announced that proved recoverable reserves could be increased by as much as 21 billion barrels by applying enhanced oil recovery (EOR) techniques to field production.[7] To do this would mean relying on a recovery of 51.5 per cent of original oil in place, rather than the 35 per cent which had been the norm in the oil industry in the 1960s, before the perfection of new technologies.[8] The government apparently based its optimism on the effect of the use of secondary and tertiary EOR in the Intisar D field, where some 68 per cent of original oil in place had been recovered. Some petroleum geologists have questioned whether EOR will be as successful in Libyan fields with structures different from that of Intisar D.[9] For this reason, most reference sources continued to list Libya's proved recoverable reserves as 22.8 billion barrels, at least through 1994.[10]

Discoveries

The first major discoveries in Libya occurred in 1959, when US companies found six large fields, including two giants, in the Sirte Basin. Esso discovered the giant Zelten field (later renamed Nasser) and Oasis, a consortium originally composed of Marathon, Continental and Amerada and later joined by Shell, discovered the Dahra and Waha fields. Esso Sirte, a subsidiary of Esso, in partnership with the Libyan American Oil Company (Arco) and W.R. Grace, discovered the Mabruk field which they decided not to develop due to technical difficulties. Amoseas, a partnership of Texaco and Socal, discovered the Beida field, and the Mobil and Gelsenberg partnership, the giant Amal field.

There were other major discoveries in the following three years. Oasis found the Defa field, Mobil and Gelsenberg the Hofra field and Esso Sirte, the Raguba field. Subsequently, the BP and Hunt partnership discovered the giant Sarir fields and Oasis discovered the giant Gialo and the smaller Samah fields. The majority of discoveries during the rest of the 1960s, all in the Sirte Basin, were smaller and less dramatic.[11] The exceptions included Amoseas discovery of Nafoora in 1966, Occidental's discovery of Idris (later renamed Intisar) and Augila in 1967 and Agip's discovery of Bu Attifel in 1968.

In order to export their oil, the oil companies which had been successful built separate, large-capacity pipelines to new terminals on the Mediterranean coast. In October 1961, Esso completed the construction of a pipeline from its Zelten field to a new terminal at Marsa Brega, a distance of 109 miles, and added a spur to this line from the Esso Sirte Raguba field. In November 1962, Oasis opened the first leg of a 267-mile pipeline connecting its several fields to a terminal at Es Sider, west of Marsa Brega. By December 1964, the Mobil and Gelsenberg partnership, working with Amoseas, had completed the 171-mile Sirtica pipeline connecting their Beida and Hofra-Ora fields to a terminal at Ras Lanuf, east of Es Sider. This Sirtica pipeline also served two small fields produced by Phillips and Pan-American (Amoco). In February 1967, the BP and Hunt partnership opened its 320-mile pipeline from the Sarir field to a new terminal at Marsa al-Hariga, next to Tobruk, well to the east of the other three terminals; Agip later shared the use of this pipeline to export its Bu Attifel production. And finally, in April 1968, Occidental completed a 132-mile pipeline from its Idris field, to a new terminal at Zueitina, some 60 miles east of Marsa Brega. This line also had a spur line to a section of the Augila field.[12] These pipelines, combined, had a total capacity of around 3.7 million b/d.

The revolution in September 1969 which brought Colonel Qaddafi to power affected exploration activities, and therefore reserve additions, badly. The majority of foreign oil companies, faced with uncertainty over their future financial arrangements, sharply reduced their operations. As Figure 5.1 shows, the number of active rigs fell dramatically after 1969 and rig activity only began to recover gradually after 1974. Subsequently, as might be expected, activity increased in response to higher oil prices and decreased as the latter fell.[13] The numbers do not reflect increased activity following the award of new exploration and production-sharing agreements (EPSA), which suggests that exploration to the present has depended very much on companies that have remained in Libya over the years and on the NOC, rather than on newcomers. Discovery results bear this

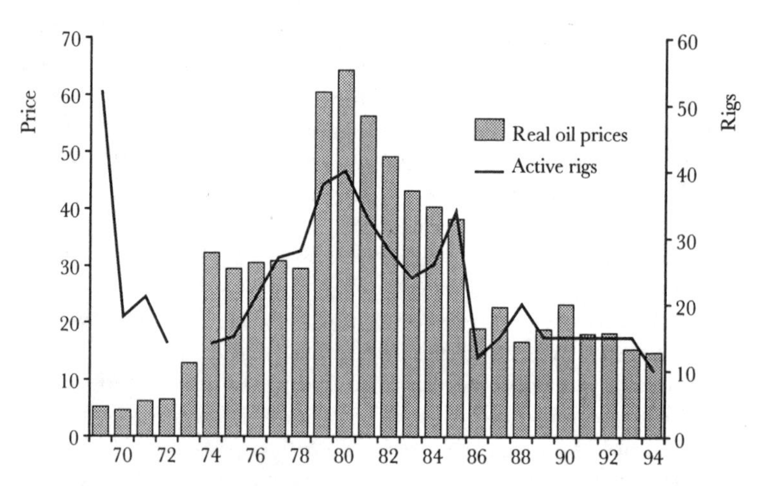

Sources: Nesté Annual Reports, Oil and Energy Trends; Waddams, The Libyan Oil Industry

Figure 5.1 Real Oil Prices and Active Rigs: Libya (1969–94)

out. There were only a few oilfields discovered in the Sirte Basin in the 25 years between 1970 and 1995, and only one of these was large. In the mid-1970s, the Arabian Gulf Exploration Company (Ageco), a subsidiary of NOC, discovered the Messlah field, with estimated reserves of 2.93 billion barrels of original oil in place, some 12 miles north-west of the Sirte Sarir field. The other discoveries were less dramatic. Occidental found a handful of new commercially viable fields in the west of the basin, Mobil found a new field near Hofra in 1971, Wintershall made a small discovery in the existing Jakhira field in 1980 and Deminex found a new field near Waha in 1983.[14] Agip considerably extended the known reserves in the Bu Attifel area.

There were, however, important discoveries outside the Sirte Basin. Rompetrol, the Romanian national oil company, and Geocom, the Bulgarian national oil company, made significant finds in the Murzuq Basin in the mid-1980s, including block NC-115 with an estimated 2 billion barrels of original oil in place. The Libyan company Ageco identified five potentially productive structures in the western Hammada al-Hamra plateau of the Ghadames Basin and began producing from at least one of these structures by the mid-1980s.

The other important area for discoveries in this period was in the Pelagian Shelf offshore Tripoli. Agip found the Bouri field there in the mid-1970s, with its estimated reserves of 2 billion barrels of original oil in place. Elf Aquitaine also found offshore oil in this area but, faced with pressure from the French government, refused to sign a contract with the Libyan government for its development.[15]

Production

There was a steady and dramatic increase in crude oil production in Libya until 1970, as Figure 5.2 shows. Output began in 1961 and reached more than 1.2 mb/d in 1965, when Esso and Oasis accounted for 88 per cent of total production. By 1967, their share had fallen to 70 per cent with the coming on stream of BP and Bunker Hunt's Sarir field and with increases from Amoseas and from the Mobil and Gelsenberg partnership's fields. In 1968, Occidental began production from its Intisar field at a rate only slightly below that of Esso and Oasis. By the

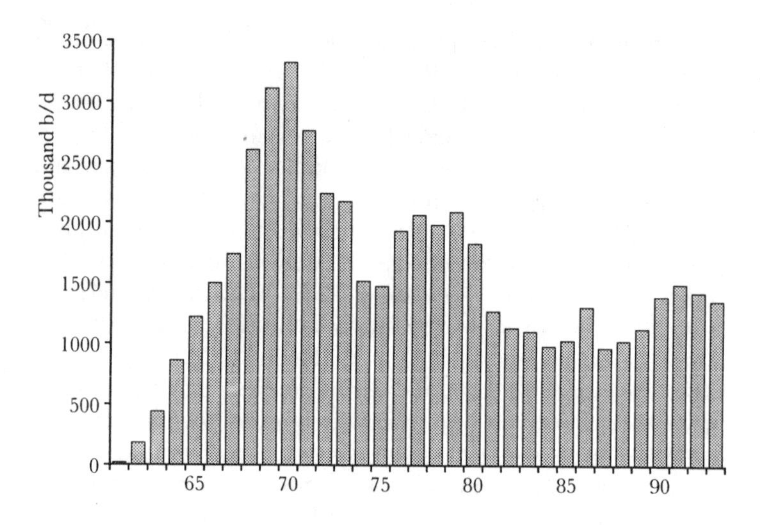

Figure 5.2 Libyan Crude Oil Production

end of the decade, Oasis led production, followed by Esso in combination with Esso Sirte, and then by Occidental. (See Appendix 5.1 for details of export crudes)

In April 1970, production peaked at 3.7 mb/d, a figure which represented the entire capacity of the pipeline systems. This suggests that the companies were determined to produce as much as possible at this time, and that they may have been doing so to the detriment of the fields involved.

The decline in production in 1971 and 1972 was the direct result of government restrictions on the production of each company. This was done partly in order to prevent over-production of fields and partly in order to force the companies to agree to higher taxes and new contractual arrangements. The first production cuts, amounting to 420,000 b/d, occurred in May 1970. Occidental, then Libya's second largest producer after Oasis, was ordered to reduce its Intisar field production by more than a third, from 797,000 b/d to 497,000 b/d. The reason given was that Occidental's failure to follow correct 'petroleum practice' was causing severe damage to oil reservoirs. The government referred to Article 11 of its 1955 Petroleum Law stipulating that a concession holder must 'diligently prosecute all his operations under the concession in a workmanlike manner and by appropriate scientific methods'. It also cited Regulation No. 8 issued in December 1968 on the conservation of petroleum resources. The reductions were stipulated on a well-by-well basis for the field.

The next oil company to have its production cut was Amoseas, the partnership of Chevron and Texaco, which was ordered in June to reduce production by 120,000 b/d, almost a third of its total, mostly from its Nafoora field. In July, the government ordered Oasis to cut production by 150,000 b/d. In August, it told the Mobil and Gelsenberg partnership to cut 40,000 b/d from its production, mostly from the Amal field, and Occidental to reduce its Augila field production by 60,000 b/d. In September, Esso was ordered to cut its production by 110,000 b/d.

The restrictions were targeted at companies which had the biggest percentage production increases during 1969. Occidental, for instance, had increased its production by almost 60 per cent in that year, and Amoseas by more than 50 per cent. Although Oasis' production had grown only slightly in 1969, it shot ahead during the first months of 1970. Both the independent companies for which Libya was the main source of crude outside North America and the majors who were badly affected by the recent Syrian closure of Tapline, the Middle East export pipeline, suffered as a result of these production restrictions. They feared that if they were unable to supply their usual customers with agreed amounts, they might lose secure markets.

The government eased its pressure to the extent of introducing penalties and credits, including a bonus system which permitted cumulative output increases of up to 15 per cent under certain circumstances. It expressed its willingness to grant increases to accommodate technical difficulties linked to low output, for instance, or as an added benefit for gas utilization programmes, or as an extra reward for new or existing industrial projects and for exploration beyond the required minimum.

Despite the protestations of the government that the cuts were only intended to protect the oilfields from overproduction, the fact that the new price agreements, which are discussed in Chapter 6, resulted in a relaxation of the government's allowable production levels, especially for Occidental, indicates that the cuts had not been entirely motivated by conservation requirements. Additional reductions in government allowables in mid-1972 were done principally to persuade companies to substitute participation agreements for their existing concession agreements, and when the participation issue was settled in late 1973, Occidental and Oasis were permitted to increase their crude liftings.

Ironically, at this point, the companies were not eager to increase their production as demand for Libyan oil had fallen and they were having difficulty finding buyers for their output. As will be seen in Chapters 6 and 7, when it succeeded in taking over the setting of the price of its crude oils, the Libyan government overestimated the value of these in competitive world markets. It failed to take into account declining tanker freight rates and a changing demand for light and heavy oils. It also ignored the fact that low-sulphur crudes were beginning to command less value in the market due to a variety of reasons.

The companies were aware of the weakening market and responded by lifting less than their overall allowable levels. The Oasis group, for instance, was first permitted and then ordered to raise its output by some 18 per cent in 1973, and failed to do so, citing problems resulting from manpower and drilling equipment shortages.

Despite the weak market, the government refused to budge from its high prices even when freight rates began to decline, making Middle East crudes less expensive in Europe. As a result of this attitude, Libya lost sales even during the period in 1973–4 when demand was increasing. Its market share among the non-communist oil-producing countries was 4.7 per cent at the beginning of 1973. By the end of the year, it had fallen to 4.2 per cent and within a few months, reached 3.8 per cent. By the end of 1974, it was only 2.8 per cent. There was little incentive for companies to increase output under these conditions and some argued that they were selling at a loss due to contractual supply obligations. Moreover, to some extent, the companies received mixed signals, as the government threatened to reduce the allowable production rates of companies which were not investing in exploration programmes.

Libyan production at the end of 1974 hovered around 800 thousand b/d, hardly more than half its level in the spring of that year. The government became alarmed, especially as it had set a target production growth rate of 7.5 per cent per annum. It realized its errors and cut prices drastically, effectively stimulating production and exports. By June 1975, output had risen to 1500 thousand b/d and exports in the third quarter of this year averaged 1900 thousand b/d.[16] As Figure 5.2 shows, Libya was able to stimulate output somewhat in the second half of this decade but never sufficiently to reach its 2 million b/d official goal.

Libyan crude oil production thus failed to fulfil the promise of the early years, when it increased phenomenally. Its annual growth from 1965 to 1970 was 22 per cent, an impressive rate even for a country beginning production. By contrast, its annual growth between 1970 and 1980 was a negative 6 per cent, compared, for instance, to a positive 10.7 per cent growth rate in Saudi Arabia.[17]

OPEC Quotas

OPEC had first considered imposing production quotas on its members as early as 1965, when it discussed a programme that envisaged pro rata production increases to accord with estimated increases in world demand. Libya, which had joined OPEC in 1962, strongly opposed the idea, insisting that it would hamper the development of its oil industry. The government said that it would not be bound by any production quota which such a programme might set, but in the event nothing came of this early incremental production planning attempt.

The plan was brought up again in the early 1980s. There had been a significant structural change in the world oil market following the oil price rises in 1973–4 and 1979–80. OPEC production fell from 29,030 thousand b/d in 1979 to 18,375 thousand b/d in 1982 as a result of a fall in demand in countries which had traditionally been the main OPEC markets due to economic recession, increased efficiency in energy usage and conservation measures. At the same time, non-OPEC oil production rose from 1979 to 1982 from 21,530 thousand b/d to 23,400 thousand b/d, with the result that OPEC's market share fell from 57 per cent to 44 per cent.[18]

OPEC responded to this threat to its production share and price level by deciding, at its conference held in March 1982, to adopt a defensive strategy based on production prorationing.

In the previous year, some OPEC members had voluntarily cut their output, but the reduced levels bore little relation to what each individual country was producing or could sell. Following its March 1982 conference, OPEC set a ceiling of 17.5 million b/d for all OPEC production, slightly below its 1982 output, and subsequently allocated national production ceilings for each individual member.

Libya's oil production was restrained after 1982 by OPEC production quotas. As an OPEC member, it was required, at least in theory, to restrict its crude oil output to the levels set by that organization in its efforts to stabilize prices. Its production quotas, listed in Table 5.2, were less than 1.5 million b/d.

Table 5.2: Average OPEC Production Quotas for Libya. Thousand Barrels/Day

1982	750
1983	1100
1984	990
1985	990
1986	990
1987	998
1988	996
1989	1037
1990	1233
1991	-*
1992	1395
1993	1390

* OPEC quotas were suspended at the time of the Iraq invasion of Kuwait in August 1990 and restored in 1992.

Sources: *Middle East Economic Digest, Arab Oil & Gas Directory 1994.*

The initial production quota of 750 thousand b/d for Libya was a considerable drop from its 1981 production of 1,220 thousand b/d. Like Iran and Venezuela, Libya refused to adhere to its allocation and continued to produce more than its due until OPEC revised its members' quotas in 1983. Libya's new limit was 1100 thousand b/d. OPEC subsequently revised the quotas in 1984, 1986 and 1988, and Libya generally produced within the limits of these, although the government protested on more than one occasion that its quota had been set too low and 'was not commensurate with Libya's historical share or ... production capacity'. It requested quota parity with Kuwait and the UAE.

Despite government protestations to the contrary, it is unlikely that Libyan production could have substantially exceeded its OPEC allowances to reach the official government production target level of 2 million b/d in this period. Factors which contributed to keeping production low included:

i. Ageing fields which were probably overproduced in the early days.
ii. Limited application of advanced enhanced oil recovery methods.
iii. Limited discovery of new reserves and delay in bringing reserves found into production partly as a result of infrastructure costs, especially pipeline construction.
iv. Departure of US companies with their financial assets and expertise.
v. Sanctions imposed by the USA and the UN which have affected equipment repairs and replacement and licensed technology.
vi. Limited government funds earmarked for investment in the oil industry. Libyan government budgets, which depend almost entirely on income from oil, have been badly affected by long-term low production levels coupled with depressed world oil prices on the one hand, and by spending on the other. The costs of major development projects (the Great Man-Made River designed to bring water from southern underground reservoirs to the coast, and the iron and steel works at Misurata) have created serious financial strains on the budget, as have military expenditures.

In 1994, Libya's oil production of 1410 thousand b/d was slightly more than that of Algeria, with 1300 thousand b/d, and slightly less than that of Nigeria, with 1880 thousand b/d.[19] It was well below that of Kuwait, with 2085 thousand b/d and bore no relation to Saudi Arabia's 8986 thousand b/d. The operating subsidiaries of NOC were responsible for the bulk of this production, as they had been since the mid-1980s.

Broken down, this output included that of the Arabian Gulf Corporation (Agoco) with a production estimated at 460,000 b/d in 1994/95 from the former BP and Hunt concessions; the Waha Oil Company with a production estimated at 440,000 b/d in 1994/95 from the Oasis concessions, with the Defa field contributing the most to this total; the Sirte Oil Company with a production estimated at 120,000 b/d to 130,000 b/d in 1994/95 from the former Esso concessions, with the Zelten field contributing the bulk of this; the Zueitina Oil Company with a production estimated at 70,000 b/d to 90,000 b/d in 1994/95 from the former Occidental concessions, mainly from the Intisar field.

In addition, four foreign oil companies were producing from several fields in the early 1990s. As noted in Chapter 4, the most important of these was Agip. Other non-Libyan producers included Veba, Wintershall and OMV. In 1995, the Total and Saga partnership had its first exports from the Mabruk field; the consortium of Repsol, Total and OMV which is developing the Murzuq field expected production to begin there in 1996 or 1997.

Potential for Production Increase

The government is relying on the use of enhanced oil recovery (EOR) techniques in older fields, as well as on the development of new fields, in order to raise oil production, and even to maintain current production levels. NOC has selected 35 reservoirs as candidates for EOR and has completed a study on resource evaluation, allocation and utilization of natural gas for secondary and tertiary recovery schemes.

Most of the major older producing fields are gas undersaturated and have active water drives. Pressure maintenance programmes were started in the first years of the extraction of oil in many of these fields in order to achieve early high production rates. Some petroleum geologists believe that this will mean that secondary and tertiary recovery programmes will not be able to substantially increase the future production of these fields.

Most of Agoco's oilfields, which include Sarir, Messlah, Nafoora, Augila and several smaller fields, were discovered in the 1960s and have water drive which augments reservoir pressure.[20] Its main Sarir C field, with 270 wells, has been producing since 1966 by natural water drive. The secondary Sarir L field, with 50 wells, has natural bottom-water drive from a large aquifer believed to be common with the other Sarir fields and with Messlah.[21] Agoco hopes to increase the recovery factor of these two fields to 47 per cent of original oil in place by the use of waterflooding. It also hopes to increase the recovery factor of the nearby Messlah field, which has been producing under diminishing natural water drive, to 42.5 per cent by waterflooding.

Agoco's other main field, the Nafoora/Augila unit, began a waterflood programme in 1969 and a secondary water injection programme in 1974. Its production peaked in 1969–70 at 300,000 b/d and in recent years has averaged only 45,000 b/d. Agoco's projected recovery factor of 41 per cent, through the use of a gas lift programme, may not result in much higher production rates. Agoco has not announced enhanced recovery programmes for its smaller fields of Bahi Amal Main, Haram, Kotla and Hamada which are now producing under natural water drives.

The Waha Oil Company, which operates its fields under a standstill agreement with Oasis, has not publicized its EOR programme. Presumably such a programme exists as there have been reports in the media of reduced natural gas pressure and consequent production falls in these fields. One report suggests that Waha is evaluating solvent flooding for some of its reservoirs.

The Sirte Oil Company also has not made public EOR programmes for the Zelten field which currently has a gas lift programme in place. By contrast, the Zueitina Oil Company has discussed its EOR experience with the several Intisar fields. The main Intisar D field underwent water flooding as soon as it started production, and within a year underwent gas injection as well. It is now undergoing tertiary recovery; some 68 per cent of its original oil in place has been recovered and Zueitina hopes to extract another 2 per cent. Sirte Oil has plans for EOR in its other Intisar fields that include gas injection, with gas from Agip's nearby Bu Attifel field. Bu Attifel currently has a water injection programme in place.

In addition to the application of EOR to older fields, the

Libyan government plans to increase output as the result of production from two new fields. One of these is block NC-115 in the Murzuq Basin in south-west Libya, said to contain between 1.5 and 2 billion barrels of proved reserves of a low sulphur crude (0.1 per cent by weight) rated at 43°API. As previously noted, this block is being developed by a consortium headed by the Spanish oil company, Repsol and the production target is 100,000 b/d. The second 'new' field is Mabruk in the central Sirte Basin which Esso discovered in 1959 but decided against developing because of anticipated high costs for producing from its badly fractured reservoir. Original oil in place is estimated at between 1 and 1.3 billion barrels of 35° to 38°API crude with a concentration of 0.26 sulphur by weight. It is being developed by Total in partnership with the Norwegian independent company, Saga Petroleum.

Libya is also hoping to extract the full potential of the Bouri field, discovered by Agip offshore Tripoli in 1976. Bouri's estimated original oil in place is 5 billion barrels of a 26°API crude with a sulphur content of 1.70 per cent by weight. It is being developed by Agip under a two-phase programme.

Lost Momentum

There was oil to be found in Libya and the oil companies discovered it quickly, once they began looking in the late 1950s. They built pipelines and coastal terminals for the export of the output of their giant fields and concentrated more on production than on further exploration at this time. Eventually, they would undoubtedly have poured more energy into securing additional fields for themselves had not the political situation in Libya and the structure of the world oil market changed so considerably. The initial momentum for the development of the Libyan oil industry was lost as a result of these changes. With a few exceptions (albeit important ones), exploration and discovery over the past two decades have been only a pale shadow of the triumphs of the early 1960s.

Production has also flagged, for a variety of reasons, including overpricing, falling demand, OPEC quotas, ageing fields and the effect of US and UN sanctions. Only a few new oilfields have been put in production since the mid-1970s and the government has not issued recent detailed field reserve and production figures. It is undoubtedly finding it increasingly difficult both to increase production from existing fields and to replace production losses with new reserves. For this reason, it is offering considerably more attractive terms to foreign oil companies whose investment of capital and expertise it needs. Many companies are convinced that there is more oil to be found in Libya, although it may not be in locations convenient to existing infrastructure or in obvious anticline reservoirs. But the strong possibility that it is there is probably enough to continue to attract oil companies to Libya, despite the disappointments to date of some recent newcomers.

Notes

1. For a description of the geology of Libya and especially the Sirte Basin see C.E. Terry and J.J. Williams, 'The Idris 'A' Biotherm and Oilfield, Sirte Basin, Libya', in P. Hepple (ed.), *The Exploration for Petroleum in Europe and North Africa* (1969), p. 33 and Louis C. Conant and Gus H. Goudarzi, 'Stratigraphic and Tectonic Framework of Libya', *Bulletin of the American Association of Petroleum Geologists*, 51/5, May 1967.
2. Exploration staff of the Arabian Gulf Oil Company, 'Geology of a Stratigraphic Giant – the Messlah Oil Field', in M.J. Salem and M.T. Busrewil (eds), *The Geology of Libya*, vol. II, (1980), pp. 523–30.
3. See John Lorenz, 'Late Jurassic–Early Cretaceous Sedimentation and Tectonics of the Murzuq Basin, Southwestern Libya', *The Geology of Libya*, vol. II, (1980), pp. 383–4 and David Thomas, 'Geology, Murzuq Oil Development Could Boost S.W. Libya Prospects', *Oil & Gas Journal*, 6 March, 1995, p. 43.
4. See Brian R. Turner, 'Palaeozoic Sedimentology of the Southeastern Part of Al Kufrah Basin, Libya: A Model for Oil Exploration', *The Geology of Libya*, vol. II, (1980), p. 372.
5. Gus Goudarzi, 'Structure – Libya', in *The Geology of Libya*, vol III, (1980), p. 891.
6. The government arrived at its reserve figures by applying a recovery factor to estimates of original oil in place in discovered fields. The recovery factor was based on what it believed could be commercially produced from these fields, using existing extraction technology and taking in account current and estimated future oil prices. To arrive at an annual proved reserves figure, it subtracted production to date from its calculation of commercial production expectations.
7. See *Middle East Economic Survey*, 33/33, 21 May, 1990, pp. A6–A7 for

a report on the Technical Symposium at which this new estimate was presented.

8. The official Libyan original oil in place reserves figure is 130 billion barrels. Applying a recovery factor of 35 per cent, gave proved recoverable reserves of 45.5 billion barrels. Using a higher recovery factor of 51.5 per cent gives proved recoverable reserves of 67.6 billion barrels. In both instances, 18.5 billion barrels, the total oil produced to the beginning of 1995, must be subtracted to calculate proved recoverable reserves at this date.
9. See, for instance, David Thomas, 'Exploration Limited since '70s in Libya's Sirte Basin', *Oil & Gas Journal*, 13 March, 1995, pp. 99–104. *World Oil* in its August 1994 issue gives a figure of proven reserves for Libya of 37.9 billion barrels which assumes more limited success with EOR techniques than the Libyan government.
10. See the *OPEC Annual Statistical Bulletin*, 1993, p. 10; *International Petroleum Encyclopedia 1994*, p. 301; *Oil & Gas Journal*, 26 December, 1994, p. 43; *BP Statistical Review of World Energy*, June 1995, p. 2.
11. The less significant Sirte fields found in this decade included Oasis' South Bahi, Esso's Jebal, Mobil's Farrud and Ora, Amoseas's Bualawn, Dor and Kotla and Elf Aquitaine's Magid.
12. The Libyan National Oil Company built the only major pipelines after 1968. These included a 298-mile line connecting the Messlah field to the Ras Lanuf pipeline and a 236-mile line connecting the Hammada al-Hamra oilfields in Ghadames to the terminal and refinery at Zawiyah, near Tripoli. A pipeline is planned to link the Murzuq NC-15 field, when it is developed, to the Hammada al-Hamra line.
13. See *Neste Annual Reports, Oil and Energy Trends*, and Frank C. Waddams, *The Libyan Oil Industry* (1980), p. 284. Waddams, who served as an adviser to the Libyan Ministry of Petroleum Affairs in the 1960s, cites as his sources the Ministry and the Central Bank of Libya.
14. Occidental's discoveries included the Almas field found in 1976; the Zella, Aswad and Sabah fields found in 1977, the Fidda field found in 1980 and the Hakim field found in 1982.
15. Its refusal was presented as a protest against Libya's interference in Chad, a former French colony.
16. J.E. Hartshorn, *Objectives of the Petroleum Exporting Countries* (1978), p. 153.
17. Kuwait's annual growth in this decade was also a negative 6.5 per cent, due to its extreme concern with the conservation of its oilfields.
18. *BP Statistical Review of World Energy*, 1987.
19. Algeria's crude includes significant condensates and natural gas liquids, so that its output in 1994 for comparison purposes to Libyan crude was probably closer to 800 thousand b/d.
20. For information on current secondary and tertiary programmes, see

Middle East Economic Survey, 31/47, 29 August, 1988, p. A4; 33/33, 21 May, 1990, p. A7; 38/4, 24 October, 1994, p. A6.

21. The smaller Sarir C-North field, however, has weak aquifer support and has been undergoing water injection.

APPENDIX 5.1

Export Crudes

The Libyan crudes traded in the international market were well established as early as 1970. These consist mainly of blends from the fields served by the five pipeline systems constructed by the oil companies in the 1960s. They include:

Amna, a blend of waxy crudes from the Amal and Nafoora fields. Its assay is generally given as 36.1°API gravity, 0.15 per cent sulphur by weight, and a pour point of 75°F. Its loading port is Ras Lanuf.

Brega, a high quality blended crude, principally from the Zelten field but often including crude from the Jebal and Ralah fields. Its assay is generally given as 39.8°API gravity, 0.2 per cent sulphur by weight and a pour point of 30°+F. Its loading point is Marsa Brega.

Bu Attifel, a very waxy crude from the Bu Attifel fields. Its assay is generally given as 43.6°API gravity, with a sulphur content of 0.034 per cent by weight and a pour point of 90°+F. Its loading port is Zueitina. It is sometimes blended with crudes from the other fields linked to the Zueitina pipeline, giving a final product with a lower wax content.

Es Sider, a blend of crudes from several fields including Waha, Gialo, Defa and Dahra. Es Sider became Libya's highest volume export crude. Its assay is generally given as 36.7°–37°API gravity, with a sulphur content of 0.27 per cent by weight and a pour point of 44+°F. It is loaded at the port of Es Sider.

Sarir, a waxy crude from the Sarir field, probably blended in the mid-1980s and subsequently with crude from the nearby Messlah field. Its assay is generally given as 37.6°–38.3°API gravity, with a sulphur content of 0.16–0.18 per cent by weight and a pour point of 77+°F. It is loaded from Marsa Hariga.

Sirtica, a high quality, light blend of crudes from a number of fields discovered by Mobil and Amoseas, such as Hofra, which are connected to the Sirtica pipeline. Its assay is generally given as 42°–43°API gravity with a sulphur content of 0.38–0.43 per cent by weight and a pour point of 77+°F. It is loaded from Ras Lanuf.

Zueitina, a light crude from the network of five fields in the Intisar system. Its assay is generally given as 41.5°API gravity, with a sulphur content of 0.28–0.31 per cent and a pour point of 45+°F. It is loaded from Zueitina. Zueitina is sometimes classified as Zueitina Blend, with a 38.4°API gravity and a sulphur content of 0.33 per cent by weight, Zueitina Premium, with a 43.7°API and 0.25 sulphur content and Zueitina Medium, with a 36.5°API and 0.16 sulphur content.

Sources: *Petroleum Intelligence Weekly's International Crude Oil Market Handbook*, June 1994, pp. G141–G150; *Oil & Gas Journal, Data Book*, 1994 Edition, 1994, pp. 343–5; Gilbert Jenkins, *Oil Economists' Handbook* (1984), p. 116.

6 MARKETING AND PRICING BEFORE THE TRIPOLI AGREEMENT

That western Europe was the natural market for Libyan crude was apparent long before the first export cargo from the Zelten field was loaded onto the *Esso Canterbury* on 12 September, 1961 for delivery to the Fawley refinery in Britain. Demand for refined products in western Europe was increasing at a prodigious rate during the build-up of the Libyan oil industry; it increased more than 200 per cent between 1961 and 1970.[1] Demand for light crudes such as those found in Libya strengthened in the latter part of this decade, reflecting a rise in demand for gasoline and for lighter distillates to be used as feedstocks by petrochemical plants in Europe.

The proximity of Europe translated into substantial transport advantages for crude exports from the North Africa coast over those from the Middle East. A tanker carrying oil from the head of the Persian Gulf through the Suez Canal had a journey of 4850 nautical miles to Marseilles and one of 6600 nautical miles to Rotterdam, and a tanker from the outlets for Saudi and Iraqi oil on the eastern Mediterranean coast had a journey to the same locations of 1600 nautical miles and 3360 nautical miles respectively. By comparison, a tanker leaving Libyan coastal terminals would need to cover only 990 nautical miles to reach Marseilles and 2700 nautical miles to reach Rotterdam. At freight rates current in the mid-1960s, North African crudes had a freight advantage over Middle East crudes of some 30 cents a barrel on the run to Rotterdam, thus allowing companies with output in both areas to net a higher price on its Libyan output.

It is not surprising, therefore, that the Libyan and Algerian share of the European market rose from 4.5 per cent in 1960 to 20.7 per cent in 1964. Production of Algerian Saharan crude, which had started modestly in 1958 at 10 thousand b/d, had risen to 330 thousand b/d in just two years. Producers of Libyan crude saw Algeria as a serious competitor in the European market, particularly in France and Spain.[2]

US quotas on imports of crude oil from sources located outside the United States enhanced the importance of the

European market for Libya. First imposed by the US government in 1957 on a voluntary basis, import quotas were made mandatory in 1959 and continued in effect throughout the 1960s. Although they were gradually relaxed towards the end, these quotas effectively closed most of the US market to new crude oil production outside North America in this decade, even for crudes produced abroad by US companies. Libyan crude oil exports to the United States averaged less than 5.5 per cent of total exports between 1963 and 1970.[3]

Libya, following the practice of other oil-producing countries throughout the 1960s, left the marketing of crude oil entirely in the hands of the oil companies which held production concessions. The exports listed in Table 6.1 resulted entirely from arrangements made by companies, except in 1971, when the Libyan National Oil Company began marketing. West Germany and Italy absorbed close to 50 per cent of all Libyan crude exports in this decade.

The major producing oil companies in Libya at this time – Esso, BP, Mobil and the Amoseas partners of Standard Oil of California (Socal) and Texaco – sold through their existing integrated networks in Europe and, where necessary, expanded these. For most of the decade, Esso accounted for almost half of all Libyan exports, with a third of its production going to West Germany and another third to Britain. The other major oil companies were exporting much less during most of these years. The Mobil/Gelsenberg partnership had strong West German marketing arrangements which absorbed as much as 75 per cent of their Libyan output; Mobil had a refinery at Bremen, Gelsenberg had one at Horst and together they owned a refinery in Bavaria. Most of the rest of their Libyan crude went to France and Britain. Half of the Amoseas partnership's production went to West Germany; its other major outlets were the Caltex Pernis refinery in the Netherlands and the Milford Haven Regent's refinery in Britain. Some was sold to the Italian Sincat refinery.

By the mid-1960s, the Oasis group of US independent oil companies (Marathon, Continental and Amerada) were accounting for 42 per cent of all Libyan crude exports. As noted in Chapter 3, these had to create their own European market networks for Libyan crude. As Marathon and Continental had both been allotted US import quotas of only 14,000 b/d, it was

Table 6.1: Libyan Crude Oil Exports by Destination 1962–1979. Thousand Barrels Per Day

	1962	*1963*	*1964*	*1965*	*1966*	*1967*	*1968*	*1969*	*1970*	*1971*	*1972*	*1973*	*1974*	*1975*	*1976*	*1977*	*1978*	*1979*
NORTH AMERICA	19	22	39	40	89	66	145	166	94	167	209	240	10	331	509	786	771	721
LATIN AMERICA	3	-	-	-	7	30	26	31	93	183	194	122	146	170	103	115	64	133
EASTERN EUROPE	-	-	-	-	4	12	9	15	1	11	66	78	9	25	54	27	50	66
WESTERN EUROPE	157	431	791	1148	1399	1620	2407	2858	3116	2368	1739	1702	1247	841	1136	999	968	1036
Austria	-	-	-	-	-	-	-	-	1	2	13	12	11	3	13	2	20	18
Belgium	19	16	27	32	57	91	83	122	128	70	26	55	30	6	5	2	3	8
France	9	34	49	115	176	221	206	334	421	348	196	121	89	54	98	78	100	114
Germany	23	145	288	462	516	406	711	751	686	508	553	497	330	285	372	332	202	297
Italy	32	59	103	122	180	354	501	662	806	657	442	566	503	294	345	315	399	356
Netherlands	18	34	71	94	146	159	187	310	302	143	104	86	13	29	45	38	46	46
Spain	0	5	40	44	58	76	157	163	147	125	56	33	55	65	95	105	108	96
Switzerland	0	6	2	12	13	17	22	30	49	45	37	33	21	7	8	2	2	7
United Kingdom	56	129	195	242	208	208	470	419	493	447	300	249	183	50	42	37	27	14
Other	-	1	17	24	42	65	64	51	58	8	2	4	5	2	0	-	-	1
MIDDLE EAST	-	-	-	-	-	-	-	-	-	-	-	-	-	-	2	-	-	1
AFRICA	-	6	26	16	5	-	1	3	-	11	16	15	14	16	2	-	-	2
ASIA AND FAR EAST	-	-	-	-	-	-	-	-	8	7	7	18	65	49	41	15	2	7
TOTAL WORLD	179	460	856	1213	1500	1717	2582	3070	3312	2747	2214	2175	1490	1431	1847	1943	1855	1966

Source: *OPEC Annual Statistical Bulletin* and *Petroleum Press Service.*

essential that they develop secure European marketing arrangements with non-affiliated refiners, and on their own, as quickly as possible. Marathon concentrated on the West German market and acquired a refinery, with Wintershall, at Mannheim. It also purchased shares in a refinery at Corunna, in Spain, and was able to supply the entire crude requirements of these two refineries from its Libyan production. In addition, Marathon bought into retail distribution systems in Italy and Germany.

Continental also focused on the West German market and purchased a 20 per cent interest in a small refinery at Karlsruhe. Much of its Libyan crude in the early days went to third parties such as DEA/Scholven and Fina as well as to ILSEA and Lombardi Petroli in Italy. In 1963, Continental decided to build a refinery in Britain to handle its Libyan oil production and its 87,500 b/d refinery at Immingham, a port on the Humber estuary in North Lincolnshire, began production in 1969. This was later upgraded to a capacity of 138,000 b/d. Earlier, it acquired a bridgehead in the UK retail market by purchasing Jet Petroleum's 600 service stations, mostly in the Midlands and North of England. It also acquired service stations in West Germany, Austria, Belgium and Italy and made firm arrangements with refineries in Italy.

By contrast, the third Oasis partner, Amerada, made little effort to acquire outlets of its own in Europe. Instead, it sold its share of the Oasis production to Continental in 1962 and 1963 and subsequently to Shell, with the crude in these deals ending up mainly in West Germany, Britain and France. After 1966, Amerada shipped its Libyan production to the refinery it had just opened at St. Croix in the Caribbean US Virgin Islands. Amerada had secured a special deal with the US government under the terms of which the crude oil processed in the refinery was not included in the US quota system, and its products could go to the United States without restriction.

Occidental, which was exporting 1 million b/d by 1970, had no refinery or retail outlets in Europe when it began producing from its Idris (later Intisar) field two years earlier. To remedy this situation, it purchased the European facilities of Signal Oil and Gas. These consisted of an 84 per cent interest in Raffinerie Belge de Pétroles with its 40,000 b/d refinery in Antwerp and a chain of about 1000 petrol stations in Britain, West Germany,

Switzerland, Belgium and the Netherlands. It also signed a long-term sales contract with the Sincat refinery in Italy, starting at 80,000 b/d and due to rise to 200,000 b/d, and a similar long-term sales contract with France's Elf Aquitaine for an average of 120,000 b/d.[4]

The Posted Price System

At the time of the enactment of the 1955 Petroleum Law, the major oil companies had full control of setting the price of the crude oil they produced from concessions in the Middle East. Theoretically, they did so on an individual basis; in fact, the nature of their interlocking equity participation and spread of interests throughout the area created a degree of interdependence.

Up until around 1960, these companies maintained a stable pricing system based on posted prices – published prices at which the posting company theoretically was willing to sell crude oil to third-party buyers at point of export. First set in the early 1950s, these posted prices were derived from a net-back formula intended to equate the price of Middle East crude oil with that of crude oil from the US Gulf delivered to New York. The delivered price of Gulf crude oil to New York, minus freight from the Middle East point of export to New York, became the posted price for Middle East crude oils, with minor variations to cover differences in the sources and quality of various Middle East crudes.

Initially, the oil companies tended to pitch their Middle East posted prices high. This was due to the fact that, for the most part, their production went directly into their own refining and retail downstream arrangements and they did not want to encourage third-party entrants. In addition, the US tax system at the time was so structured that companies producing in the Middle East and paying taxes to host governments on the basis of posted prices were eligible for considerable foreign tax credits from the US Treasury.

Sales of Middle East crude oils to third parties did occur, however, and the actual prices realized in these sales began to fall well below the level of posted prices as a result of increased supply from new and existing sources of production. Realized market prices declined nearly 20 per cent in some markets between 1957 and 1959.[5] They continued to fall in the European market throughout the 1960s, with an interruption in late 1967 and early 1968 as a result of the Arab–Israeli War. As a result, companies were obliged to offer substantial discounts from their posted prices in sales of their crude oil to others in Europe. Discounts from postings were not published, but they were generally known in the trade to be as much as 40 cents per barrel in 1959 and 1960, when the posted price for the standard Middle East crude, 34°API Arabian Gulf oil was around $1.80 a barrel.[6]

As the discrepancy between posted prices and market prices increased, the major oil companies found that they needed to lower posted prices, on which taxes were based, to maintain their profit levels. Governments in countries where oil was being produced, however, strongly opposed any attempts by oil companies to lower the posted prices which served as the basis for their calculations of royalties and other taxes, since they would suffer a loss of revenue if this was done. Esso's declaration in August 1960 that it was reducing the posted prices for Middle East crudes crystallized the views of these governments and led directly to the formation of OPEC.

The major oil companies found it increasingly difficult to maintain their former control over Middle East crude oil prices by the end of the 1960s. Not only did they face the increasing political power of OPEC but they had to contend with growing competition in markets from independent oil companies, especially those operating in Libya, and from newly formed national oil companies. Their guaranteed access to oil reserves was diminished and their ability to balance supply and demand, and therefore to maintain set prices, was threatened. It was effectively ended in 1973, when they agreed, in order to stave off nationalization, to producer countries' participation in their operations.

Libyan Posted Prices

The 1955 Libyan Petroleum Law specified realized price, not posted price, as the basis for the calculation of payments due to the government for oil when it was produced. It described this as 'the average free competitive price'.[7] Realized prices in early 1961, however, when Esso was preparing to begin the first exports of Libyan oil, were low compared to posted prices and this led the government to change its mind on the basis for tax calculations. A Royal Decree in July 1961 specified posted price as the basis for the assessment of taxable income and royalty. It defined posted price as:

> the price f.o.b. Seaboard Terminal for Libyan crude oil of the gravity and quality concerned arrived at by reference to free market prices for individual sales of full cargoes and in accordance with the procedure to be agreed between the concessions holder and the commission; or if there is no free market for commercial sales of full cargoes of Libyan crude oil then posted price shall mean a fair price fixed by Agreement between the concession holder and the commission or in default of agreement by arbitration having regard to the posted price of crude oil of similar quality and gravity in other free markets with necessary adjustments for freight and insurance.[8]

This description echoed the definition of posted price given in the 1952 concession agreement between the Iraqi government and the Iraq Petroleum Company and its affiliates.

The oil companies working in Libya anticipated this change. As Table 6.2 shows, in 1961, before the Royal Decree was announced, Esso posted a price for Libyan crude of $2.21/barrel for 39°API crude oil, with a decrease of 2 cents/barrel for each degree of gravity below 39° and a ceiling of $2.23 for crude oil with 40°API or higher. Esso arrived at this figure, which included a 40 cents/barrel freight premium for Libyan crude, by averaging the posted prices of four Middle East crudes from Saudi Arabia, Iran and Iraq. The only other Libyan producer to post a price at the time was Oasis, which based its posting on that of Esso, adjusted for a lower gravity crude. This was effectively a hypothetical posting since, as noted in Chapter 3, Oasis persuaded the government of its need to discount heavily in order to sell in European markets where it lacked the integrated refinery connections of the majors. The government allowed realized, not posted prices, to serve as the basis for the assessment of taxes on Oasis and later, Occidental – a decision

Table 6.2: Posted Prices, 1955–71

Year	*Arabian Light $/Barrel*	*Esso for Libyan Crude $/Barrel*
1955	1.93	
1956	1.93	
1957	2.08	
1958	2.08	
1959	1.90	
1960	1.76	
1961	1.80	2.21 (39°API)
1962	1.80	2.21
1963	1.80	2.21
1964	1.80	2.21
1965	1.80	2.21
1966	1.80	2.21
1967	1.80	2.21
1968	1.80	2.23 (40°API)
1969	1.80	2.23
1970	1.80	2.53
1971	2.29	3.45

which led to antagonism between the independent and the major oil companies as well as to problems for the Libyan government with OPEC.

The government objected strenuously to the Esso posting as at least 30 to 40 cents/barrel too low. It argued that the price did not take into account the lightness of Libyan crudes, which therefore yielded more gasoline and kerosene than heavier crudes. It also argued that the low sulphur content of Libyan oil had been ignored.[9] Esso countered that the value of low-sulphur crude was not great and that the advantage of this quality of the crude oil from its Libyan fields was offset by its high wax content. And finally, the government insisted that Esso had not used the correct adjustment for freight charges in its calculations as it had based these on spot market charges for single voyages rather than the Average Freight Rate Assessment (AFRA) set by an independent panel.

In addition to specifying posted prices as the basis for the assessment of taxes in 1961, the government also described the procedure for fixing posted prices. Unfortunately, it did so in a manner sufficiently ambiguous to lead to different interpretations by the government and the oil companies. It stipulated that:

> the concession-holder or its affiliates shall from time to time establish and publish its posted price for Libyan crude oil petroleum of the gravity and quality concerned, which shall be the price at which crude petroleum of that gravity and quality is offered for sale by the concession holder or its affiliates to buyers generally in cargo lots f.o.b. Seaboard Terminal.[10]

This passage was interpreted by the oil companies to mean that a concession-holder could post prices unilaterally. The government argued that the posted price definition in the July 1961 Royal Decree explicitly required the agreement of the Petroleum Commission to any price posted.

The disagreement on this point was raised again when Mobil started producing from its Amal field towards the middle of 1966 and posted a price of $2.10/barrel for 36°API crude oil, 5 cents below the Esso posting, adjusted for gravity allowances. Mobil claimed that this lower posting was necessary because of the high wax content of Amal crude which gave it a high pour point. The Amoseas partners, Texaco and Socal, announced the same posting at this time for their Nafoora crude, which also had a high wax content. At the end of 1966, BP posted its Sarir oil at a price 2 cents lower than the Mobil and Amoseas postings, citing higher transport costs in addition to high wax content. The government formally protested all these postings.

Market prices for crude oil in Europe soared in the first few months after the outbreak of the Arab–Israeli war in June 1967 as a result of the closing of the Suez Canal and the brief embargo by some members of the Arab League of oil sales to countries believed to be assisting Israel. When Libyan shipments resumed after the embargo, the high freight rates of a longer journey of Arabian Gulf oil around the Cape of Good Hope due to the closure of the Suez Canal increased the freight advantage of Libyan oil over Arabian Gulf oil to northwest Europe from about 40 cents/barrel to at least $1.10/barrel.[11]

In August 1967, the Libyan minister of petroleum affairs sent a letter to the oil companies requesting them to submit proposals for an increase in posted prices. This letter cited the rise in market prices of crude oil and products in the European market, the favourable geographical location of Libya for this market following the closure of the Suez Canal and the rise in freight rates to Europe for Middle East crude oils. It did not mention the quality advantages of Libyan oil or gravity allowances. Talks between the government and the companies began that month, with the oil companies refusing to increase posted prices on the grounds that high market prices were the result of unusual conditions and would fall again.[12] Instead, they proposed, and implemented, a temporary elimination of the allowances against taxes which had been agreed at the time of the royalty expensing deal in 1964.

The following month, OPEC passed a resolution at its Thirteenth Conference in which it noted that it had:

> heard the statement of the Head of the Libyan Delegation concerning the unjustifiable basis on which Libyan posted prices have been calculated since their inception, and also on Libya's rightful claim for a separate upward adjustment of posted prices f.o.b. Libyan ports in the light of current tanker freight rates.[13]

The resolution went on to say that Libya's claim was 'justified by prevailing economic conditions in the international oil industry' and that OPEC therefore resolved 'to support any appropriate measures taken by the Libyan Government to safeguard its legitimate interests'.

Although it publicly supported the Libyan government's position, the focus of most OPEC members at this time was on the elimination of the 6.5 per cent discount allowance off posted prices which had been granted to the oil companies to alleviate their tax obligations when they accepted royalty expensing. Saudi Arabia, Iran, Kuwait, Abu Dhabi and Qatar accepted an offer from the companies regarding the phasing out of this allowance. Libya flatly refused the offer, pointing out that it did not address the substantial tax discount given to the companies for light crude production. As Libyan production at this time consisted entirely of crudes over 37°API, this discount reduced government tax income.

In January 1968, Occidental posted a price of $2.23, based

on the Esso posted price, for crude oil from its recently opened Idris and Augila fields. The government protested immediately, reiterating its standing objections to the Esso posting and stressing that Occidental's crude oil gravity of 42° had not been taken into account.

The companies continued to refuse to compromise on the issue of posted prices, particularly as market prices began to fall, and the government continued to press its demands, albeit ineffectually.

Collision Course

Not long after it seized power in September 1969, the Qaddafi government took an aggressive line in pricing, as it did in ownership and production. Its radical approach to pricing derived partly from its nature as a revolutionary regime and partly from the position it was trying to establish *vis-à-vis* other Arab governments, especially members of OPEC. There were also strong internal reasons for taking a bold stand on prices arising from the impasse it had inherited from the previous regime. Frustration had built up regarding the level of posted prices set by the oil companies just as it had earlier over the practice followed by the independent oil companies of discounting very considerable marketing expenses in their declarations of net taxable income. Disputes over these two issues had kept the heat on in discussions between the Idris government and the companies throughout the previous decade. The eventual curbing of the independent companies' expense deductions by a cooperative effort between the major oil companies and the Idris government in 1965, described in Chapter 3, set a precedent regarding effective tactics for the Qaddafi government in its negotiations over pricing. It was able to profit from the ill-feeling resulting from long years of pre-tax expense deductions by Oasis and Occidental assuming, it turned out correctly, that the majors would not willingly come to the aid of the independents if they were under pressure. Probably the most significant factor in the unfolding of the drama of pricing in the early 1970s, however, was the youth and confidence of the individuals in the new Libyan revolutionary government.

The government became aware that demand for oil products in western Europe was increasing rapidly at this time. In addition, it noted a growing demand for low-sulphur fuel oil in the USA following the enactment in 1966 of the US Environmental Protection Act which restricted the use of high-sulphur oil in the generation of electricity. These two factors had combined to result in higher prices and profits for oil companies producing low-sulphur Libyan crudes. In December 1969, four months after it came to power, the Qaddafi government established a committee to negotiate an increase in posted prices with the oil companies. It pointed out that although government revenue had increased in recent years as a result of the expensing of royalties and other tax concessions made by the companies, the basis for posted prices had not been altered since Esso's initial 1961 posting. Negotiations between the committee and the companies began early in 1970, but little progress was made.

In May 1970, the European crude oil and product markets experienced a mini-shock when telephone engineers ran a bulldozer through the Trans-Arabian Pipeline (Tapline) on Syrian territory. As this pipeline normally delivered some 500,000 b/d of Saudi Arabian crude oil to the Sidon terminal on the Mediterranean, an immediate supply shortage resulted when Syria refused to repair the damage.[14] Tanker rates and product prices soared as did the price of crude oil landed in Europe.

Later that month, the Libyan government inaugurated its programme of reducing the production allowances for individual oil companies described in Chapter 5. At the same time, it signed an agreement with Algeria pledging joint cooperation and coordination in oil matters, especially in efforts to achieve higher posted, tax reference prices.[15] The Libyan authorities had been following events in Algeria closely. In 1968, Algeria had persuaded Getty Petroleum to raise its tax reference price and in June 1970, it nationalized the interests held by Shell, Phillips, the West German company Elwerath-Sofrapel and the Italian company AMIF when these refused to accept terms similar to those in the Getty agreement. In July, Algeria announced a new posted price for French companies – effective retroactively to 1 January, 1969 – of $2.85/barrel, an increase of 80 cents over the existing price. Subsequently, in February 1971, it seized 51 per cent of all French interests in the country.

Posted Prices Raised

The capitulation of the oil companies to Libyan government demands for higher posted prices, and thus to increased tax obligations, is a story which has been often told.[16] The government concentrated on the independent companies which were extremely dependent on Libyan crude oil supplies and would need access to other sources to meet their market commitments if they resisted the government. These independents controlled more than 50 per cent of Libyan production at this time, a fact which encouraged the government to believe that, if necessary, it could survive a boycott of production by the major companies if it had succeeded in getting agreement from the independents to higher prices.

On 1 September, 1970, Occidental capitulated to government demands, apparently after having its request for assistance in supplying oil to its customers refused by Esso. It agreed to an increase in its posted price of 30 cents, from $2.23/barrel to $2.53/barrel, to be increased by a further 2 cents annually for five years, reaching $2.63/barrel in 1975.[17]

Within six weeks, the rest of the oil companies working in Libya, with one exception, had also come to an agreement with the government over posted prices. The Oasis partners of Continental, Marathon and Amerada Hess signed an agreement in the third week of September; Shell held back until mid-October. Gelsenberg, Hunt, Liamco and Grace agreed at the beginning of October as did the Amoseas partners, Socal and Texaco. The following week, Esso, Mobil, BP and Elf Aquitaine accepted. Phillips, however, refused and surrendered its small field which, in fact, was not breaking even. The new base gravity was 40°API, with a 2 cent differential increase for each degree of gravity above 40° and a 1.5 cent differential decrease for each degree below 40°.[18] The previous 5 cent penalty on waxy crudes and the 2 cent freight differential for Hunt and BP's Sarir crude were eliminated. In addition to increased posted prices, the tax rate for the companies was raised from 50 per cent to an average of over 54 per cent.

The major oil companies were concerned about the effect of these settlements regarding posted prices on other OPEC members. They also saw the tax rate increase resulting from increased posted prices – although Libya described this as a method for recouping retroactive debts – as a possible threat to their arrangements elsewhere. In addition, they were worried that if countries which did not have retroactive debts were successful in getting tax rate increases, Libya would demand further tax hikes. Their concerns were not without foundation. Several producing countries demanded higher posted prices and a 55 per cent tax rate; in December 1970, Venezuela passed a law raising its tax rate to 60 per cent and claimed the right to raise oil prices unilaterally without reference to the companies. At its meeting in Caracas later that month, OPEC passed a resolution to consolidate the gains made by its members regarding rates of taxation, levels of posted prices and gravity differentials. It called for a minimum 55 per cent tax rate.

The effect of the continued Tapline closure and cutbacks in Libyan output to prevent overproduction, kept oil prices high in Europe. In January 1971, the Libyan government announced that the agreements signed the previous autumn did not reflect current conditions. It argued that the market price of crude oils had risen to such an extent that the agreed increase in posted prices absorbed only a small amount of the extra revenue derived by the companies from their sales. It presented the companies with a 'non-negotiable' demand for a further 5 per cent rise in tax rates and another posted price rise.

The companies decided – too late – to collaborate to maintain their control of prices. With the exception of the two state-owned companies, they all signed the so-called Libyan Producers Agreement early in 1971, committing themselves to act in concert *vis-à-vis* the Libyan government and to assist each other when necessary.[19] They agreed to refuse unilateral tax rises. In addition, they agreed to set up a safety net of supply-sharing, promising to make good the loss of oil for any company placed under production restrictions by providing it with crude from their Libyan production at cost, or from other production if Libyan oil was not available. The amount supplied would be 100 per cent of cutbacks in 1971, 80 per cent in 1972 and 60 per cent in 1973. The test of the Libyan Producers Agreement was made public only in 1974, when Bunker Hunt filed a claim for damages against Mobil for breaking it.

Teheran Agreement

At its December 1970 meeting in Caracas, OPEC had agreed that its members should pursue negotiations on taxes, prices and differentials with the oil companies. The latter orchestrated their reply through its London Policy Group. This recently established forum in which companies could discuss strategy and tactics received special dispensation from the US Justice Department with regard to violation of US anti-trust laws. In January 1971, the companies sent a public 'Letter to OPEC' signed by 24 companies, including all the members of the Libyan Producers Agreement. This proposed an 'all-embracing negotiation' for the achievement of an 'overall and durable settlement'. Iran, however, opposed negotiations involving all OPEC members, as did the North African producers, and the talks in Teheran were limited to producing countries from the Gulf, with neither Libya nor Algeria participating.

The Teheran Agreement reached on 14 February, 1971 incorporated the substance of the OPEC Caracas Resolution and the promise of no further changes for five years. The tax rate was set at 55 per cent and posted prices were raised by 35 cents/barrel with subsequent annual increases of 5 cents/barrel plus 2.5 per cent until 1975 in order to counter inflation. The allowances dating from the royalty-expensing agreement were eliminated and the price differentials for different gravities were those proposed in the Caracas Resolution. These were 1.5 cents/barrel for crude oil for each degree of gravity below 40°API and 2 cents/barrel for each degree of gravity above 40°API.

Tripoli Agreement

Negotiations between the oil companies and the Libyan government took place in Tripoli during and after the talks in Teheran. The Libyan government was dissatisfied with the Teheran Agreement on grounds of what it saw as insufficient price increases and an inadequate premium for short-haul freight. The Libyan National Oil Company had recently started to sell royalty oil on its own and the government argued that since it was able to sell oil at substantially higher prices than the posted prices, the latter were far too low. The other OPEC members with outlets on the Mediterranean – Iraq, Saudi Arabia, Algeria – supported Libya in these negotiations.

The Tripoli Agreement of 20 March, 1971 required each oil company in Libya to amend its Deeds of Concession to cover the agreed changes. These included a stabilized tax rate of 55 per cent except for Occidental which had a further 5 per cent liability. The posted price for 40°API crude oil was raised to $3.32/barrel, a figure derived from a base posting of $3.07, a Suez Canal Allowance of 12 cents as long as the Canal remained closed and a temporary freight premium of 13 cents. The latter was to be adjusted to changing AFRA and World Scale freight rates. The gravity differentials embodied in the Teheran

Table 6.3: Libyan 40°API Crude Oil Price April 1971. Dollars Per Barrel

Base Posting Price	
Posted price prior to Tripoli Agreement	2.550
General increase of Teheran Agreement	0.350
Low sulphur differential (a)	0.100
Fixed freight differential	0.050
Increase 2 April, 1971	0.020
	3.070
Permanent Posted Price Change	
Base Posting Price	3.070
Inflation premium (b)	0.127
	3.197
Temporary Elements	
Suez Canal allowance	0.120
Freight premium (c)	0.130
Final Price	3.447

(a) Applicable to crude oils with 0.5 weight per cent sulphur content or less.
(b) 2.5 per cent of the base posting of $3.07 plus 5 cents.
(c) To be determined quarterly on the basis of 0.0058 cents/barrel for each 0.1 percentage point of World Scale by which the assessed LR 2 AFRA exceeds World Scale 72.

Source: Gabi Jarjour, *OPEC and Oil Pricing Structure: Analysis of OPEC Official Resolutions* (1982), p. 17.

Agreement were accepted. A low-sulphur premium of 10 cents was agreed where sulphur content was less than 0.5 per cent by weight and this was to be increased by 2 cents annually up to 1975. The base posting, as in the Teheran Agreement, was also to increase annually up to 1975. The agreement signed by each company contained a commitment to undertake further exploration in Libya.

Libya's posted price for 40°API rose to $3.45/barrel in March 1971, in accordance with the terms of the Tripoli Agreement. The calculations for arriving at this price are shown in Table 6.3. At the time, the price of what became the official OPEC marker crude, Arabian Light, as seen in Table 6.2, was $2.29/barrel. Libya's price was thus $1.16/barrel more than that of the Gulf producers, compared with a difference of $0.73 after the September 1970 agreement with Occidental and $0.43 in the years before this. By 1974, the difference had increased to $4.12/barrel. Not surprisingly, Libya lost the cost advantage which it had over Middle East oil in markets in Europe and North America as a result of its pricing position at this time.

Notes

1. *OPEC Annual Statistical Bulletins*, 1966, 1970, 1975.
2. The feeling was mutual. According to the Algerian government, 'France always considered Libyan crude a direct and important competitor of Algerian oil and used it as the only basis of reference for fixing the fiscal terms applicable in Algeria'. Algerian Government, *Background Information on the Relationship between Algeria and the French Oil Companies* (1971), p. 9. Libyan oil production outstripped that of Algeria by 1964 and, by 1967, was more than double. *Petroleum Press Service*, October 1965, p. 365; *British Petroleum Statistical Review*, 1967, p. 19.
3. Its experience of relying on European markets helped Libya to weather the complete loss of the US market in the mid-1980s without much damage.
4. Both of these contracts ran into trouble in the early 1970s. The Elf Aquitaine agreement ended in early 1973 as a result of increases in Libyan crude prices. The Sincat agreement was also affected badly by price considerations. See *Middle East Economic Survey*, XII/44, 28 August, 1969, p. 4; *Petroleum Intelligence Weekly*, 25 December, 1972, p. 4.
5. Albert L. Danielsen, *The Evolution of OPEC* (1982), p. 135.

6. Ian Skeet, *OPEC: Twenty-Five Years of Prices and Politics* (1988), p. 4.
7. Shukri Mohammed Ghanem, *The Pricing of Libyan Crude Oil* (1975), p. 241.
8. Ibid, p. 51
9. A premium for low sulphur content only came into being after legislation was introduced in the USA and Japan to control the burning of high-sulphur fuels in industrial and domestic furnaces. The 1971 Tripoli Agreement introduced a sulphur premium in Libyan postings, initially at 10 cents/barrel and rising over the following three years to 25 cents/barrel.
10. Kingdom of Libya, Ministry of Petroleum Affairs, *Libyan Oil 1954-1967* (1968), p. 42.
11. Frank C. Waddams, *The Libyan Oil Industry* (1980), p.162.
12. Prices did come down in the European market during 1968 and 1969, despite the continued closure of the Suez Canal, but did not reach their previous levels.
13. Supplement, *Middle East Economic Survey*, X/49, 6 October, 1967, p. 1.
14. Tapline remained closed until early 1971 when relations between Syria and Saudi Arabia improved.
15. Algeria had long been negotiating with France without success over the revision of the agreements which regulated the terms under which French oil companies could exploit the oil deposits of the Sahara. It claimed, as did Libya, that the correct posted price for North African crudes should have been $2.65, not $2.21/barrel.
16. See, for instance, 'The International Oil "Debacle" Since 1971', *Petroleum Intelligence Weekly Supplement*, 22 April, 1979, pp. 1–36, which contains an account by G. Henry M. Schuler, of Nelson Bunker Hunt.
17. In addition, Occidental's tax rate rose from 50 per cent to 58 per cent, with the company's commitment to pay 5 per cent of pre-tax profits towards the development of the Kufrah oasis no longer applicable.
18. Previously, the gravity differential had been subject to a ceiling of 40.9°API, with a 2 cent differential decrease for each degree below this ceiling.
19. The 17 companies which signed included Amerada Hess, Atlantic Richfield, BP, Bunker Hunt, Continental, Exxon, Gelsenberg, Grace, Gulf, Hispanoil, Marathon, Mobil, Murphy, Occidental, Shell, Socal and Texaco. The American companies received an assurance from the US government that it would not take action against them for violation of US anti-trust laws by forming this agreement. See Waddams (1980), p. 238.

7 MARKETING AND PRICING AFTER THE TRIPOLI AGREEMENT

The Idris government intended that the Libyan national oil company (Lipetco), which it established in 1968, should actively market royalty crude oil received by the government in lieu of cash payments and also its share of production when it entered into partnership with foreign oil companies. Nothing was done to implement this plan, however, until a year after the Qaddafi regime assumed power in September 1969.

When the Libyan National Oil Company (NOC) which replaced Lipetco did begin marketing crude oil for its own account in the autumn of 1970, it followed, in some respects, the marketing pattern established by the foreign oil companies in the previous decade. Like the independents (the Oasis partners and Occidental) it depended first on third-party sales and then, eventually, sought to secure its own wholesale and retail outlets in Europe. Except for a brief period in the late 1970s when there was a surge of sales to the United States, Libyan marketing efforts concentrated on Europe. Reflecting a long-standing historical connection, Italy increasingly became the main recipient of Libyan exports. This trend was intensified after 1983, with the purchase of shares in the Italian downstream company Tamoil, which had a refinery at Cremona in Italy and later one at Collombey in Switzerland, as well as a network of retail outlets (see Chapter 8 for further details). West Germany remained an important market partly because of the purchase of the Holborn refinery from Coastal Oil. On the whole, however, as will be seen in the next chapter, Libya's acquisition of an integrated European network was opportunistic, at least in the beginning, and lacked a masterplan.

Libya's marketing policy deviated from that followed by the independent oil companies, however, when it entered into barter deals with the Soviet Union, Bulgaria and Romania in the spring of 1972, beginning a long-term policy of offering crude oil in barter and debt repayment agreements to countries in the communist bloc (although often the crude in these deals ended up physically in west European markets as a result of swap arrangements). These barter and debt repayment deals reflected the intrusion of political factors in the commercial marketing of oil. The linking of sales contracts to promises to participate in Libyan development plans is another example of this type of intrusion which has been more pronounced in downstream than in upstream activities.

Brega Petroleum Marketing Company

In July 1970, the government announced the setting up of the Brega Petroleum Marketing Company, a subsidiary of NOC, to market Libyan crude oil abroad. The only source of crude oil for Brega at this time came from the government exercising its option, spelled out in concession contracts, to take all or part of its royalty in kind rather than in cash. After 1973, it received crude oil from its production share in participation agreements with foreign oil companies.

Brega made its policy regarding the European market clear at the very beginning. It announced that it was willing to entertain bids for annual renewable contracts for crude purchases from oil companies, independent refiners and marketers but that applications from companies without established refining and marketing facilities in Europe would not be considered. It added that it would give preference to bidders willing to participate in Libyan development plans, which included projects in the agricultural and industrial sectors as well as in the oil sector. Successful bidders were expected to adhere to government destination embargoes, if and when these occurred.

Brega's first sale of 300,000 tons of royalty oil was made in late 1970 to OMV, the Austrian state oil company. This deal involved small daily shipments, over a period of four months, of Oasis royalty crude. Although the agreement was renewable in 1974, OMV let it lapse. Brega arranged another direct sale at this time involving the sale of Occidental and Sarir royalty crude to Egypt. Most of this shipment ended in Rotterdam in a swap arrangement because the Egyptian refinery in Alexandria could not handle crude with a high wax content.

In December 1970, Libya decided to begin trading on the international oil market. Brega made an agreement with Wetco, a London-based Egyptian trading company, to sell a large proportion of Oasis royalty crude to independent European refiners over the coming year. A month later, the arrangement with Wetco was increased to include the total offtake of royalty crude due to the government from Oasis to the end of 1971. Brega also arranged for the sale of royalty crude from Esso, the BP and Bunker Hunt partnership, Amoseas and the Mobil and Gelsenberg partnership to a Swiss brokerage group, Berola S.A., which marketed mainly in West Germany, Italy and the Netherlands. In the end, most of the intended Wetco and Berola sales did not take place because of disputes over prices and Brega sent most of the royalty crude intended for sale by these agents to the Sincat refinery in Italy. Some small consignments of crude oil also went to Bulgaria and Romania in 1971.[1]

Marketing of Sarir Crude

Since accepting royalty in kind rather than in cash was an option and not a requirement, Brega was not obliged to take and sell large quantities of crude in its first 18 months. In October 1970, NOC acquired production of its own when Phillips Petroleum decided to surrender its concession rather than agree to new tax and price terms. What was involved was a small, unprofitable field, however, and its acquisition was of little consequence in terms of government crude supplies. Under these circumstances, Brega was able to proceed slowly in its marketing and selling efforts. It preferred slow involvement into the business, with the pace determined by market conditions and by the need to gain management experience.

The government's abrupt nationalization of BP's share of the Sarir field at the end of 1971, however, put Brega under great pressure to succeed as a marketing organization. The Arabian Gulf Oil Exporting Company (Ageco), the NOC subsidiary created to take over the BP interest in Sarir, acquired not only BP's half share of Sarir production but also a sizeable portion of Bunker Hunt's share which BP, as operator, had been taking and selling.[2]

Brega's efforts to sell Libyan crude abroad up to this point had been largely unsuccessful. The high prices which resulted from the government's negotiations with the oil companies in late 1970, and the terms of the spring 1971 Tripoli Agreement, made Libyan crudes overpriced in the European market compared to Middle East crudes. Moreover, price stability did not follow in the wake of the Teheran and Tripoli Agreements. The US suspension of the convertibility of US dollars into gold in August 1971 led to a devaluation of about 8.57 per cent in the value of the dollar against most major currencies of the world. As virtually all crude oil and product sales, including those of Libya, were denominated in dollars, OPEC called for new negotiations with the oil companies regarding the effect of the devaluation. The resulting so-called First Geneva Agreement in January 1972 raised posted prices in the countries which were parties to the agreement by 8.49 per cent. Libya, which had not been a party, accepted this settlement five months later.

Following the BP nationalization, Brega was faced not only with a lack of successful experience and a market less interested in Libyan crude oil but by the fact that Sarir crude was not easy to market as many refiners could not cope with its high wax content. The situation was made even more difficult after the British government contacted the governments of all major oil-importing countries requesting them to prevent purchases of Sarir crude by oil companies and traders under their jurisdiction. French and German oil companies assured the British that they did not intend to buy any of BP's share of Sarir crude, and BP claimed an affirmative response from all western nations to its request for support.

The Libyan government tried to ease the situation by reducing the amount of Sarir crude that Brega had to sell. It ordered the field's production drastically cut and asked the remaining partner, Bunker Hunt, for help in marketing, a request that was refused.[3] In the first two months following the takeover, Brega was only able to sell three or four cargoes of Sarir crude. One went to Sicily for refining where it remained in storage during a court case brought by BP claiming ownership of the shipment, and another went to the Aegean.

Brega finally found success with the Soviet Union which wanted to import crude oil for re-export to the west European market in exchange for hard currency.[4] In March 1972, the Soviet oil export agency Sojuznefteexport (SNE) signed an agreement containing a renewal option with Brega for the purchase of Sarir crude over one year at a rate of 40,000–

45,000 b/d. The agreement formed part of a five-year barter deal involving cooperation in oil exploration and refining and the construction of a nuclear power plant. It also covered Soviet assistance in the construction of desalination plants, an iron and steel factory and railways. There was no indication whether the Sarir crude was intended for the Soviet market or whether SNE would place it in eastern Europe or elsewhere as the agreement gave the agency the right to re-export it. Some, at least, was sent to the Black Sea port of Odessa which had a large export refinery.[5] The British government formally protested in vain to the Soviet government that this agreement was a violation of international law. During 1972 and 1973, Brega managed to obtain orders for a total of 5.5 million tons of Sarir crude from countries in eastern Europe through barter deals, adding Yugoslavia, Bulgaria and Romania as customers. Some of these deals came unstuck. Yugoslavia originally agreed to take 20,000 b/d, but the quantity was reduced after a disagreement over price.[6] Crude exports to Romania were suspended in the first year as a result of a disagreement over Romania's fulfilment of a housing construction contract and only the agreement with Bulgaria went ahead as planned. Eventually, shipments to eastern Europe became more regular. As Table 7.1 shows, Libyan crude exports to eastern Europe continued into the 1990s, averaging some 4 per cent of Libyan crude exports annually; in 1986, however, they surpassed 16 per cent of total exports.

By the spring of 1973, aided by a crude oil shortage in Europe, Brega managed to sell Sarir crude outside the Soviet and eastern European bloc. More European buyers were forthcoming after a Sicilian court rejected BP's claim to ownership of the crude shipped by Brega to Sicily for refining by Sincat. The trade press in mid-1973 reported that Brega was selling between 210,000 b/d and 245,000 b/d of Sarir crude to buyers in Europe, the United States and South America, as well as to east European 'anchormen' in June 1973.[7] Known purchasers included two US companies, Coastal and Ashland Oil. The latter had a three-year agreement which involved the processing of Sarir crude in Italy and the subsequent export of gasoline to the US East Coast in a trade-off for West Coast gasoline. Some 40,000 b/d went to the John Latsis Petrola Hellas Elefsis export

Table 7.1: Libyan Crude Oil Exports by Destination (1980–1993). Thousand Barrels Per Day

	1980	*1981*	*1982*	*1983*	*1984*	*1985*	*1986*	*1987*	*1988*	*1989*	*1990*	*1991*	*1992*	*1993*
NORTH AMERICA	599	299	32	5	4	5	-	-	-	-	-	-	-	-
LATIN AMERICA	132	32	19	14	8	7	5	-	-	-	-	-	-	-
EASTERN EUROPE	50	48	42	33	51	55	175	52	50	28	85	50	45	35
WESTERN EUROPE	817	593	841	860	827	787	850	745	825	836	994	1130	1095	1045
Austria	22	13	23	14	8	8	30	32	32	20	29	10	19	16
Belgium	4	2	60	47	31	1	35	48	29	25	23	37	29	38
France	38	25	49	64	121	69	46	36	66	46	58	78	63	43
Germany	311	173	219	218	116	110	95	129	202	217	232	248	233	232
Italy	251	176	214	219	258	285	297	271	312	366	489	522	515	487
Netherlands	12	12	30	81	57	60	35	16	7	12	7	9	9	2
Spain	86	80	81	74	77	80	133	77	81	79	108	110	130	106
Switzerland	23	10	19	37	18	15	14	31	27	9	23	38	32	28
United Kingdom	-	5	39	19	9	9	6	1	-	-	5	-	-	8
Other	-	-	-	-	-	-	-	-	-	-	-	-	-	-
MIDDLE EAST	69	65	28	18	17	15	16	6	8	4	-	-	-	-
AFRICA	2	-	8	5	1	6	6	4	4	2	5	20	20	15
ASIA AND FAR EAST	25	26	5	2	22	19	15	3	3	2	6	20	20	15
TOTAL WORLD	1693	1063	974	937	930	895	1067	810	890	872	1090	1220	1180	1110

Source: *OPEC Annual Statistical Bulletin* and *Petroleum Press Service.*

refinery in Greece and Brazil's Petrobras agreed to take 25,000 to 30,000 b/d over ten years. Brega had even more Sarir crude to sell in the second half of 1973 after the nationalization of Bunker Hunt.[8]

Chuffed by its success in placing Sarir crude, Brega expanded its operations. In July 1973, it took over from ENI the marketing of NOC's production allotment from the Bu Attifel field in which the two companies shared an interest. Shortly thereafter, it arranged the sale of Oasis and Occidental royalty crude to Sohio, a midwest US refiner and marketer of gasoline. Earlier, the government announced its intention to take control of at least some shipping arrangements and embarked on an ambitious plan for a fleet of Libyan tankers.[9] Orders totalling $217 million were placed with shipbuilders in Japan and Sweden for ten tankers of varied deadweight tonnage, with delivery scheduled for 1976. In addition, the government purchased a 50,000 dwt Norwegian-flag tanker for the transport of products from the Sincat refinery in Sicily to Libya, and ordered two smaller product tankers from a Yugoslav shipyard.[10]

1973 Upheavals

Brega's expansion of activities came at a time when the world oil market was unstable for a number of reasons. The US government's decision to abolish the import quota system in favour of import fees was one cause. Another was growing tension in the Middle East between the Arab states and Israel. The currency exchange markets were in a turmoil following the second devaluation of the dollar in February 1973 which signalled the final end to the Bretton Woods exchange rate system.[11] And finally, inflation was going out of control. The Teheran Agreement contained an accommodation for inflation of 2.5 per cent; by 1973, the general level of inflation was at least 7 to 8 per cent.

Oil companies were increasing their profit margins, in spite of higher taxation rates, as a result of market price increases, and spot prices for sales by OPEC national oil companies (including that of Libya), exceeded the level of posted prices. OPEC members felt that they were being short-changed and those from the Gulf States began negotiations with the oil companies over revisions of the terms of the Teheran Agreement on 8 October, 1973, two days after the outbreak of hostilities between Israel and Egypt. On 12 October, the oil companies asked for an adjournment of the talks. On 16 October, the Gulf States OPEC negotiators issued a communiqué stating that they had decided to set the posted prices of crudes in the Gulf without reference to the oil companies. These posted prices would be based on 'actual market prices' and calculated with regard to the posted price of Arabian Light crude oil. The communiqué indicated that individual nations could determine their sulphur premiums and that the Geneva Agreements were to continue in force.

Under the managed pricing system which evolved after this date, OPEC set a reference 'marker' price, that of Arabian Light crude oil, which could only be adjusted at OPEC meetings. It then attempted to administer a price structure of official selling prices for all OPEC members by sanctioning differential values from the reference marker to account for differences in gravity, sulphur content and location in terms of markets. After 1973, posted prices served for tax reference only, determining what a government would receive from its participation and production-sharing agreements regardless of official and spot market prices. Libya's posted price remained higher than its official selling price although it reduced its posted price when necessary to restore the margins of companies on their equity crude entitlements when there had been a reduction in official market sales prices.

The marker price that OPEC set in October 1973 became unrealistic as a result of the outbreak of the Arab–Israeli war in that month. Production cut-backs and embargoes imposed by the Arab states on oil shipments to the USA and the Netherlands raised the price of oil to new heights. Spot prices exceeded the official selling prices of many producers.

In December 1973, Brega decided to stage an auction for Libyan crude. Several other OPEC members, including Iran, Kuwait, Abu Dhabi and Qatar, were also holding auctions at this time at which they were making sales at levels higher than their current official selling prices. The estimated 700,000 b/d of crude under offer by Brega at this auction consisted of royalty payments in kind, output from the nationalized Sarir field and

NOC's share in partnership agreements which would ordinarily have been buy-back oil for the companies.[12] Brega reported that 40 to 60 bids were received at the auction although it was rumoured that some of these were from 'phantom' companies. Few bidders expressed a willingness to participate in Libyan development schemes as the price for qualifying for crude shipments. Although the government did not issue information on the successful bidders and the prices at which deals were arranged, it would appear that the auction was not a success. Even deals that were made later fell through. One reason for this failure was the fact that some producing companies had not agreed to all the terms of participation demanded by the government and were threatening legal action against buyers of their crude on the grounds that the government's right to sell it was in dispute.[13]

At the beginning of 1974, the price of the official OPEC 34°API marker crude oil, Arabian Light, had reached $11.65. As Table 7.2 shows, the Libyan government's official selling price at this time for its marker crude, Es Sider, was $16.00, some $4.35 higher than that of Arabian Light. This price included very generous freight allowances and other differentials, including an increase of the low-sulphur premium to $1.33/barrel from its previous $0.14/barrel.

The Libyan price was far too high at the beginning of 1974 and became increasingly out of line with prevailing conditions as the year progressed. The government insisted on continuing the embargo which it had imposed, along with other Arab states during the October 1973 Arab–Israeli war, on exports of crude to the USA and to Caribbean refineries supplying the USA. In doing so, it ignored the value of Libyan low-sulphur crude for the US market and the fact that other OPEC members were now cultivating this market, having lifted their embargoes early in the year. The government also ignored important market signs. It overlooked the collapse in spot freight rates, a change in demand for light and heavy products, increased excess capacity in desulphurization plants and a relaxation of sulphur regulations in certain countries.

As a result of these oversights, Libya's market share among the non-communist oil-producing countries fell from 4.7 per cent at the beginning of 1973 to 2.8 per cent by the end of

Table 7.2: OPEC Marker Crude and Libyan Government Official Selling Price Es Sider Crude. US Dollars

	OPEC Marker	*Es Sider*	*Differential*
Start 1974	11.65	16.00	4.35
End 1974	11.25	12.50	1.25
Start 1975	10.46	11.86	1.40
End 1975	11.51	12.21	0.70
Start 1976	11.51	12.21	0.70
End 1976	11.51	12.40	0.89
Start 1977	12.09	13.74	1.65
End 1977	12.70	14.00	1.30
Start 1978	12.70	13.80	1.10
End 1978	12.70	13.68	0.98
Start 1979	13.34	14.52	1.18
End 1979	24.00	29.78	5.78
Start 1980	26.00	34.50	8.50
End 1980	32.00	36.78	4.78
Start 1981	34.00	40.78	6.78
End 1981	34.00	37.28	3.28
Start 1982	34.00	36.50	2.50
End 1982	32.00	35.15	3.15
Start 1983	29.00	30.15	1.15
End 1983	29.00	30.15	1.15

Sources: Ian Skeet, *OPEC: Twenty-five Years of Prices and Politics* (1988); *International Crude Oil and Product Prices*, 1994; Gilbert Jenkins, *Oil Economists' Handbook*, 1984.

1974. As Table 7.3 shows, Libya was unable to sell its crude oils at the prices it had set and production therefore plummeted. The fall in 1974 output compared to 1973 was massive, particularly for NOC which then accounted for the largest single volume of crude exports.[14] Production from the nationalized Sarir field fell 42 per cent and from the nationalized Amoseas fields, 61 per cent. Companies also suffered from the effect of mandated high prices and threatened to reduce or even cease production unless these were lowered. One company was reported as offering to sell some of its crude on the spot market at prices below cost in an attempt to bolster its oil exports in the face of sagging demand.[15]

Table 7.3: Libyan Crude Oil Production and Prices

Year	*Quarter*	*Production Thousand Barrels/Day*	*Official Selling Price Es Sider Dollars*
1974	1st	1879	16.00
1974	2nd	1761	14.60
1974	3rd	1374	13.20
1974	4th	1032	12.50
1975	1st	960	11.86–11.70
1975	2nd	1277	11.40
1975	3rd	1996	11.10
1975	4th	1793	12.21
1976	lst	1793	12.21
1976	2nd	1931	12.21
1976	3rd	1954	12.40

Sources: *Petroleum Economist*, November 1976, p. 420; *International Crude Oil and Product Prices*, 1976.

As the extent of the fall in production became apparent in 1974, the government began adjusting its prices. In April, it made its first reduction followed by others in May and October, in some cases setting a flat rate and in others, a variable price scale taking the gravity and wax content of its seven crudes into account.[16] Despite these measures, sales were not forthcoming and production continued to fall. This was because the repeated price reductions were not sufficient to restore the appropriate price differential with the OPEC marker. Libyan oil production almost halved between the first quarter of 1974 and the first quarter of 1975 despite a price reduction of $4.30/barrel in the same period. It was only when a further price cut of 30 cents per barrel was made that production recovered, doubling in just one quarter. This illustrates an important point: in a competitive oil market it is essential for a country to set the price differential between its own crude and substitutes at the correct level. In other words, adjustments are about selling at the correct levels and not only about moving in the right direction.

At the beginning of 1975 the government lifted the embargo on sales to the USA and again lowered its price. By the middle of 1975 the tide had turned. The government was not convinced, however, and it lowered prices further, a reduction which may have been unnecessary. By October, it felt sufficiently secure to impose the general OPEC price increase of 10 per cent. In July 1976 it added a further modest increase as a result of active demand, but it was careful subsequently not to raise its official price above the spot market level. By the third quarter of 1976, production had reached the government target of close to 2 million b/d.

As well as reducing its posted price, the Libyan government sought to increase exports through bilateral agreements with European governments for crude sales. In 1974, it signed a bilateral agreement with the French government which listed a number of economic and industrial projects that France and Libya could undertake, promising France 'certain quantities' of Libyan oil. It also signed a bilateral accord with Italy at this time in which Libyan crude was exchanged for 'economic, technical and industrial facilities and other products of which Libya might have need'. A day after signing this accord, the government granted Agip, the subsidiary of ENI, new offshore acreage. A subsequent long-term bilateral oil and trade agreement allowed Agip to double its direct offtake of Libyan crude in return for Italian goods and services. In 1975, the government signed a five-year crude supply agreement with the Spanish government as part of an overall exchange and cooperation package.

After 1975, the government was less bullish in its price setting although, as Table 7.2 shows, it continued to maintain differentials and consistently priced its marker Es Sider crude higher than the OPEC marker – but the difference was greater at some times than others.[17] It also reacted more quickly to market changes than the official OPEC adjustment system. In general, Libyan prices followed the pattern of those in other OPEC countries, stabilizing in 1975 and then rising gradually up to the beginning of 1979. In the price rise resulting from the Iranian Revolution, Libya's Es Sider crude peaked at $40.78/barrel at the beginning of 1981. It then declined in a more-or-less regular fashion to a new, relatively stable level of $30.15/barrel in 1983.

Shift to the Brent Market

By the mid-1980s, the crude oil market was experiencing a relentless fall in demand at the same time that non-OPEC production was increasing. In September 1985, Saudi Arabia decided to abandon the OPEC pricing system for which it had served as linchpin for more than a decade and introduced netback pricing for its export sales. The OPEC pricing system had never been very successful. Market conditions had exerted a strong influence on the demand for different crudes and the producer governments that set differential levels often misinterpreted market signs. Another difficulty was that the posted price of an OPEC member, which was related to the OPEC reference marker price, also served a tax reference role, determining what its government received in taxes from oil companies. Producer governments therefore had other reasons than market conditions to determine the level at which they set their posted prices.

Saudi Arabia's decision to abandon the OPEC pricing system was followed by an oil price collapse and a permanent transformation of oil pricing methods worldwide. At first, there was considerable confusion regarding posted and contract prices. The situation in Libya was particularly difficult due to the decision of the US government in early 1986 to order all US companies to cease their Libyan operations.[18] When prices finally began to stabilize towards the end of the year, OPEC had abandoned its system of administered oil pricing and was concentrating on a strategy of production volume restraint. Its Middle East and African members increasingly began to draw up formulas which related their prices to those of North Sea Brent or Dubai crudes. The 15-day forward, so-called Dated Brent market was then well established, having emerged through the development of spot trading in North Sea crudes in the late

1970s as an informal market unregulated by official authorities, functioning in accordance with the law of contracts with voluntarily accepted standard procedures.[19] Following the collapse of the OPEC pricing mechanism, Dated Brent became the marker crude for price determination for many of the world's crude oils. Even Saudi Arabia was pricing contract sales by using a discount off Dated Brent by the end of 1988.

In 1987, Libya adopted a price formula for its Es Sider crude based on Dated Brent minus 50–60 cents, adjusting the price of its other crudes to this figure in accordance with their gravities. It then shifted, in November 1988, to a quasi-netback formula, and for several months based its prices half on Dated Brent plus or minus an adjustment factor for each type of crude and half on a product netback formula.[20] By mid-1989, it had returned to a formula based solely on Dated Brent plus or minus adjustment factors to which it was still adhering in the mid-1990s.

Table 7.4 shows Libya's Es Sider discount from the Dated Brent price. From the beginning of 1987 until the beginning of

Table 7.4: Differentials, Es Sider and Dated Brent. Monthly Averages in US Cents

January 1987	-40	January 1992	-46
June 1987	-45	June 1992	-67
December 1987	-83	December 1992	-52
January 1988	-82	January 1993	-52
June 1988	-53	June 1993	-57
December 1988	-25	December 1993	-54
January 1989	-40	January 1994	-50
June 1989	-45	June 1994	-35
December 1989	-52	December 1994	Parity
January 1990	-63	January 1995	Parity
June 1990	-63	December 1995	-05
January 1991	-66		
June 1991	-74		
December 1991	-50		

Source: *Platt's, Middle East Economic Survey.*

1994 this averaged minus 55 cents. It then went to parity with Dated Brent and in the early months of 1995 ranged from 10 to 20 cents above Dated Brent, reverting to a slight discount later in the year. The prices of Libya's other crudes are related to that of Es Sider, corrected for gravity and sulphur content, except in the case of offshore Bouri crude oil whose price was set in terms of Iranian Heavy Oil when it began to be traded.

Towards the end of 1990, Libya made some changes in its price formula, instituting a floor price for all Libyan light crudes. Contract sales were adjusted on a cargo-by-cargo basis in line with any narrowing of the differential between Es Sider and Dated Brent, or Bouri crude and Iranian Heavy, as reported in trade publications. Prices were f.o.b, calculated on the basis of the average price of Dated Brent, or Iranian Heavy, five days around the Bill of Lading, plus or minus adjustment factors, with payments due 30 days from the Bill of Lading.

Recent Marketing Policies

Brega's emergence as an international presence on world crude oil markets was precipitated by the exodus of the American companies from Libya in 1986. Its policies at this time and subsequently have several characteristics:

A continued focus on Italian and West German markets. Although the volume of exports has changed, roughly reflecting changes in production volumes, the export destination pattern has remained substantially the same since 1962, as a comparison of Tables 6.1 and 7.1 indicates. Despite the surge in exports to North America, mainly the United States, after 1975, the main market for Libyan crude oil has remained unquestionably western Europe. Its share of Libyan crude oil exports averaged 93 per cent between 1963 and 1970, decreased during the next decade, and returned to 90 per cent between 1982 and 1993. If the unusual sale in 1986 of 16.4 per cent of exports to eastern Europe is excluded, the average west European share since 1982 is more than 91 per cent.

Within western Europe, the West German share has declined somewhat. It was about 27 per cent in the first decade or so of Libyan production, dropped to 15 per cent when exports to the

United States surged, and averaged 22 to 25 per cent from 1982 to 1993, except in the mid-1980s. By contrast, Italy's share increased from an average of 26 per cent in the late 1960s to almost 44 per cent in the late 1980s and early 1990s. The US market was significant in the late 1970s and then decreased, ending in 1986. Britain's market share was 15 per cent until 1971; as the North Sea came more and more into production, its share fell to practically nothing.

In the late 1980s and early 1990s, Libya is reported to have sold significant quantities of crude oil to the Soviet Union. These sales do not show up in trade statistics as being to the Soviet Union since they were used to fulfil Soviet supply contracts with companies in Italy, Germany, France and other eastern European countries. Some of these sales were handled by Moscow, and others by Finland's Neste Oy trading outlet. On the other hand, while continuing its traditional focus on western Europe, abetted by exports to eastern Europe and the former Soviet Union republics, Libya has tried to diversify its markets in recent years, seeking new customers and more flexibility in its types of sales agreements. By the early 1990s, it was selling some crude in the Far East through Korean companies, including the occasional cargo to China.

Preference for state oil company buyers. In 1992, more than 60 per cent of Libya's total crude exports went to state oil companies and to its affiliated European downstream outlets of Tamoil and Holborn. Customers included not only state oil companies in eastern European countries but those in western Europe as well, such as Agip and OMV.

Linking sales contracts to exploration commitments. This policy first became prominent in the early 1970s, often connected to deals with state oil companies. When there was a seller's market in late 1979, Brega announced that it would give the largest share of sales contracts to state-backed firms and other companies willing to sign up for exploration. In order to emphasize this point, it cancelled and cut back several supply contracts to third-party customers and offered crude with better terms to companies already participating in exploration. A later example of such linking with state oil companies is the evergreen contract for crude sales with Spain's Repsol combined with an exploration and production-sharing contract for valuable acreage in the Murzuq basin.

Continued barter trade. Barter deals with the former Soviet Union and eastern European countries have continued into the 1990s, although not at the level they reached in the previous decade. Barter deals with the former Soviet Union have been important for both parties. In the earlier years, they served as a means for Libya to acquire military hardware and, occasionally, engineering services. As noted earlier, Libyan crude imports, re-exported to west European markets, provided a valuable source of foreign exchange for the Soviet Union. They also facilitated the Soviet Union's provision of oil to its satellites, especially Bulgaria and East Germany. Libyan oil shipments, however, were insufficient to cover the cost of imports and Libya's debt to the Soviet Union, and later to Russia, was reported to have reached $2.4 billion by the early 1990s.[21] In mid-1995, the two countries signed a series of economic and trade agreements worth around $1.5 billion, apparently after having resolved the issue of debt repayment.

Data on the amount of Libyan oil exported to the Soviet Union are hard to find. Soviet imports of crude oil from OPEC members in the first half of the 1970s, which included imports from Iraq and Algeria as well as from Libya, have been reported as varying between 60,000 b/d and 227,000 b/d.[22] Imports from Libya alone between 1972 and 1990 are shown in Table 7.5.

Libya has also made barter agreements with other countries. It signed a trade pact with Egypt which included the barter of oil for rice, sugar, cement and other goods; one with Uganda involving a swap of crude and products for tea, coffee and beans; and one with Sudan involving an exchange of oil and other products in return for camels, meat and animal fodder. In 1993, the Chinese oil trader Sinochem was supplying Croatia with Libyan crude under a barter deal.

Debt repayment agreements. In a sense, debt repayment deals are a form of barter trade with the buyer having provided goods and services beforehand. The Libyan government has often used

Table 7.5: USSR Crude Oil Imports from Libya. Barrels Per Day

Year	Barrels Per Day
1972	33,500
1973	37,500
1977	21,000
1980	34,000
1982	119,000
1983	118,000
1984	125,000
1985	96,000
1986	92,000
1987	52,000
1988	102,000
1989	20,000
1990	20,500

Sources: *Soviet Oil, Gas and Energy Databook*, 1978; *Middle East Economic Digest*, 15 October, 1990; *CIS and East European Energy Databook*, 1991, 1995.

crude oil as a means of paying off debts to some of the many foreign contractors involved in Libyan construction projects. These debts were estimated by the *Financial Times* in the mid-1980s as totalling $4 billion, with the largest amount owed to Italian firms. Other creditor countries whose contractors were owed large sums of money at this time for finished projects included Turkey, South Korea, West Germany, Japan, France, Spain and Greece.[23] Later important creditors have been the Soviet Union/Russia and Korea.

In some cases, Libya made blanket repayment agreements involving crude oil exports directly with governments. For instance, it signed a government-to-government agreement with Italy covering the settlement of commercial debts by means of crude oil deliveries. One country which has long been taking Libyan crude oil in repayment for contractors' work in Libya is Turkey; another is India, which built two airports in Libya in the 1980s. In the early 1990s, Libya was making debt repayments in the form of crude and products exports to Korea for Korean companies involved in the construction of the Great Man-Made River.

Avoidance of spot market sales. By the late 1980s, with its combination of downstream processing acquisitions in western Europe, long-term contracts to traditional west European customers and barter agreements to the Soviet Union and eastern Europe, Libya could fine-tune sales to suit its needs without recourse to the spot market.

Its political isolation, exaggerated by US and UN sanctions, is undoubtedly one of the reasons why Libya prefers secure contract sales to the spot market. NOC depends heavily on regularly renewed contract sales to Tamoil, OMV, Agip, Repsol, Elf Aquitaine and others for the disposing abroad of the crude it produces. The fact that customers accept and renew long-term contracts suggests that the NOC is following good business practices, including reasonable contract negotiations and terms and reliable delivery. Conservative behaviour in business dealings by Libyan officials seems an anomaly in the face of the radical image which the Libyan government presents to the outside world and which is echoed in the Western media. In fact, it is because of this wild reputation that NOC must be seen to be acting with extreme propriety; otherwise it would be unable to attract and keep the customers it needs.

When a shipment of Libyan crude does appear on the spot market, it is usually oil company equity crude. The sellers of Libyan crude in the spot market in the early 1990s have been mainly Agip, Veba and OMV although Libya's European downstream affiliate Tamoil sometimes sold through spot market transactions.[24] Es Sider, Libya's quality crude, is traded rarely in this market; most deals involved Zueitina, followed by Sirtica and Sarir blends. Table 7.6 shows that the frequency of spot markets trades was highest in 1987 and gradually decreased in the years that followed to almost nothing in 1993.

Up until 1988, spot prices were very close to Es Sider official prices, with adjustments for gravity. After 1989, there was considerably more variation in these adjustments to Dated Brent than there was for official posted price adjustments, and occasionally involved prices higher than Dated Brent.

Reorganization

In October 1990, the government dissolved Brega Marketing and transferred its operations to a department within NOC. This move allowed NOC to control export destinations and

Table 7.6: Spot Trading in Libyan Crudes

Year	*Number of Trades*
1986	17
1987	60
1988	48
1989	38
1990	36
1991	19
1992	9
1993	7

Source: *Petroleum Argus.*

sales arrangements in a more systematic way in the face of the threat of increased UN sanctions. It resulted in an even greater reliance on contract sales. In 1992, the new marketing department tightened its control over sales by issuing rules regarding restricted destinations in sales contracts. At the end of 1994, Libya again altered its price formula for term contracts so that the price formulas for light crudes were calculated on the average of Dated Brent for the whole month in which the cargoes in question were lifted, plus or minus differential adjustment factors.[25] This change was said to have been made after discussions between the Libyan NOC and its term customers. Although the altered system coincided with a marked increase in the volatility of Dated Brent, the main reason given by the government for the change was to simplify NOC's financial administration of prices.[26] On the other hand, it meant that customers could pay different amounts for the same crude within the same month.

Although the Qaddafi government had been strikingly successful in its aggressive marketing and policies in its first few years and had successfully translated the attraction of the location and quality of Libyan oil to its own advantage, its lack of experience became critical in 1974 when market conditions changed. Sales of Libyan oil fell dramatically as a result of the government badly misreading market signs and insisting on too high a price for Libyan crude. In the two decades that followed this sobering lesson, Libyan marketing and pricing policies have been cautious and conservative. The government has preferred secure arrangements – contract sales, barter agreements, state oil company customers – to exposure in the spot market. It has complied closely to the prices set by OPEC and when the OPEC pricing mechanism collapsed, pegged its prices to the Brent Market. It has also devoted much effort and attention, as will be seen in Chapter 8, to developing its own refining outlets both in western Europe and in Libya.

Notes

1. Foreign oil companies working in Libya occasionally sold to east European markets. Gelsenberg sold Amal crude to Romania in 1969 and Occidental had a standing sales contract with Yugoslavia which involved shipping Libyan crude oil to a west European destination for swapping.
2. Naming the subsidiary Arabian Gulf was a gesture intended to show that the islands whose loss to Iran precipitated the BP nationalization were situated in what Libya and other Arab nations felt should be properly termed the 'Arabian Gulf' and not the 'Persian Gulf'.
3. Bunker Hunt had no effective marketing outlets in Europe and had depended on BP to market its equity share of Sarir production.
4. For a discussion of Soviet policy regarding oil imports, see Margaret Chadwick, David Long and Machiko Nissanke, *Soviet Oil Exports: Trade Adjustments, Refining Constraints and Market Behaviour* (1987), pp. 119–23.
5. See 'The International Oil "Debacle" Since 1971', *Petroleum Intelligence Weekly*, Special Supplement, 22 April, 1974, p. 35.
6. Another difficulty was that Yugoslavia's refineries were not geared to handle waxy crudes in colder weather and its ports were not equipped with heated lines. *Petroleum Intelligence Weekly*, 27 November, 1972 p. 4.
7. *Petroleum Intelligence Weekly*, 25 June, 1973, p. 2.
8. The Libyan government settled compensation for nationalization with British Petroleum in November 1974, and with Bunker Hunt in September 1975.
9. The idea was not new. As early as 1966, the Idris government had included in concession agreements a promise that oil companies 'grant priority' to Libyan-flag vessels, and noted that it expected to enforce this promise when Libya had a tanker fleet. See *Petroleum Intelligence Weekly*, 11 June, 1972, p. 4.
10. By 1990, Libya had 12 crude oil carriers and four product tankers under the administration of its General National Maritime Transport Company. The fleet's tonnage stood at 1,650,000 dwt, making it the

largest fleet in Africa and the fourth largest among OPEC members at this time.

11. The Second Geneva Agreement in June 1973 following this devaluation made another adjustment to posted prices to which Libya agreed.
12. The Oasis partners and Occidental had buy-back contracts with the Libyan government following their acceptance of participation. In the beginning, these were one-year contracts; later they became contracts for six months and, by the early 1980s, for three months.
13. See *Petroleum Economist*, February 1974, p. 51.
14. If the participation oil which was sold back to the oil companies was included in the tally, NOC sales exceeded 55 per cent of total production in 1974.
15. *Petroleum Intelligence Weekly*, 7 April, 1975, p. 3.
16. One cause of low production in late 1974 was the dispute of the government with Esso over the prices at which the latter was selling LNG. When the government refused to allow Esso to export LNG at existing prices, it ordered Esso's oil output to be halted from its associated fields in order to prevent flaring of unused gas.
17. In 1977, for instance, it raised its prices as the USA began to come out of recession and its imports increased. However, it did not raise them as much as Nigeria or Algeria did. See J.E. Hartshorn, *Objectives of the Petroleum Exporting Countries* (1978), p. 153.
18. Analysis has shown that the cut-off from the US market did not result in a statistically significant reduction of prices for the sale of Libyan crudes elsewhere. See Paul Horsnell and Robert Mabro, *Oil Markets and Prices: The Brent Market and the Formation of World Oil Prices* (1993), pp. 261–2.
19. Ibid, pp. 73–82.
20. This was made up from prices of premium gasoline (20 per cent), jet kerosene (11 per cent), gas oil (25 per cent), low-sulphur fuel oil (36 per cent), and naphtha 4 per cent, less a fee for processing and freight.
21. *Middle East Economic Survey*, 38/44, 31 July, 1995, p. A13.
22. Jeremy Russell, *Energy as a Factor in Soviet Foreign Policy* (1978), p. 24.
23. *Middle East Economic Survey*, 7 October, 1985, pp. A7–A8.
24. Source data assembled by *Petroleum Argus*.
25. The price formula for heavy offshore Bouri was calculated in a similar fashion, based on the average price of Iranian Heavy.
26. *Platt's Oilgram Price Report*, 73/6, 1 October, 1995, pp. 1, 4.

8 REFINING IN LIBYA AND IN EUROPE

Although foreign oil companies were in total control of the Libyan oil industry in the early years, the government wanted a role in the downstream operations which it expected to be built in Libya. In an interview with the *Middle East Economic Survey* in the summer of 1966, Oil Minister Fuad al-Kabazi announced that 'what we are aiming for are complete arrangements for participation in the profits of all phases of oil industry operations'.[1] The operative word here was 'profits'; there was little expression by the government of a desire to share in the management of refining or distribution activities.

In order to encourage companies to develop the downstream side of the business in Libya, the government inserted a requirement in several concession contracts that the signatories construct refineries in Libya in exchange for exploration and production acreage. In only one instance, however, as the result of an agreement with Esso, did the government succeed in achieving this.

In later years, the Qaddafi government entered into the business of constructing and managing domestic refineries and eventually became involved in the ownership of refineries and distribution systems in Europe. But whereas commercial considerations based on the requirements for developing an oil industry can be seen to be major determinants of upstream Libyan energy policies over the years, downstream decisions involved other criteria. The decision, for instance, to focus on the production of heavy oil to the exclusion of light distillates in refineries built in Libya was made mainly so as to facilitate an ambitious development plan for electricity generation using fuel oil. The corollary of this decision, reliance on European refineries for light distillates by providing these with regular deliveries of Libyan crude has never been profitable for the oil industry. Moreover, the inability to produce gasoline within Libya became a serious problem as demand for this product increased.

Although there were sound commercial and other reasons for Libya to acquire downstream facilities abroad, it is difficult to identify the criteria the government used in its selection of specific European refineries for purchase. These choices appear to have been made on an opportunistic, *ad hoc* basis without the benefit of a calculated, long-term downstream programme based on supply and demand calculations. There are a number of reasons why OPEC member countries, especially those who were accumulating petrodollars for which there was little real opportunity for investment at home, might choose to buy into refining and retailing outlets abroad. With their shortage of arable land and water, with virtually no other natural resources besides oil, and with only small populations, it is not surprising that Kuwait and Libya – and later Saudi Arabia – should look abroad for investment opportunities.

There were commercial as well as financial reasons for buying into European refining for some oil-producing countries. These were particularly important for Venezuela, for instance, which inaugurated a very active overseas investment policy even though it did not lack domestic investment possibilities. What motivated Venezuela, like Kuwait, was a desire to secure outlets for poor quality crudes in a market which was becoming increasingly less interested in these varieties as a result of tighter environmental regulations regarding emissions. For Libya, with its light, low-sulphur crudes, the quality of its production was not a problem.

The desire to secure market outlets in the late 1970s and 1980s, however, was not confined to oil-producing countries with poor crudes. Not only was world oil demand faltering in this period, but OPEC was faced with increased competition in European markets from non-OPEC supply, particularly from the North Sea. Libya shared the common wish of OPEC member countries to protect its production volume when demand was weak by having its own, more-or-less dedicated, European refinery and retail outlets.

An important reason why Libya decided to buy into downstream Europe was undoubtedly its loss of the US market and the threat of ever more restrictive measures by the US government. This had acquired what appeared to be an irreversible momentum. In May 1978, the US government had banned the export of military hardware, including aircraft, to Libya. In December 1981, President Reagan had issued an executive order requesting all US citizens to leave Libya as soon as possible. In

February 1982, the US government had banned the importation of Libyan oil into the United States and had placed restrictions on US exports to Libya except for food and other farm products, medicines and medical supplies. It had also denied the export to Libya of US oil and gas technology and equipment which was not readily available from sources outside the United States. In 1983, the US government had asked other nations to support curbs on exports to Libya; two years later, it was to ban the importation of refined petroleum products from Libya.

Another reason for securing its own downstream outlets in Europe – which Libya shared with Kuwait and Venezuela – was the somewhat unfounded belief that if and when oil prices fell, a company which had its own refineries and retail distribution systems could expect that downstream profits would compensate, to some extent, for upstream losses. There was also the unquestioned advantage of providing opportunities to gain expertise in refining and marketing. And there was an unexpected financial bonus in the late 1980s for oil producers that owned European refineries as these generated revenues in hard European currencies which were appreciating against the dollar, thereby providing them with a hedge against the dollar's decline. Sometimes tax considerations were important – the Venezuelan oil company PdVSA, for instance, wanted a source of income that its government could not tax easily.

What provided the opportunity for oil producers to invest in west European downstream properties was the wish of the majors, especially US companies, to dispose of their European refineries in the early 1980s. Most attractively, they were willing to sell these cheaply due to an overabundance of such facilities. They had constructed much new refining capacity in western Europe after the Second World War on the back of strong growth in demand for products and developments in refining technologies. West European governments had encouraged them in this as they saw refineries as a way of reducing their growing need for imported oil products which created foreign exchange problems. Refinery construction in western Europe acquired its own momentum and capacity continued to increase despite the decrease in demand for oil products which followed the major oil price rises of 1973 and 1979. Capacity utilization rates fell substantially from around 90 per cent in 1970 to just under 70 per cent in the early 1980s and profitability of refinery activities declined significantly. In parts of western Europe, utilization fell to around 55 per cent in 1981. To make matters worse, new taxes on gasoline further diminished the growth rate of demand for this product which had already fallen as a result of higher prices and lower economic growth. Many major companies were anxious to shed their loss-making facilities at almost any price.

Libyan Refineries

Libya's first thoughts were to install refinery capacity within its borders in order to meet domestic product demand. Originally, it hoped this would occur in the natural course of the development of the industry and the government included a clause in its 1955 Petroleum Law giving any concession holder who discovered oil the right to construct and operate a refinery in Libya. When no company seemed inclined to consider refining in the country, the government then decided to apply pressure to this end. The opportunity to do so arose when two small oil companies, Liamco and Grace, applied for permission to give Esso a 50 per cent interest in their concessions. The government agreed to this request on condition that, if oil was discovered, Esso would construct a refinery in Libya to supply the internal market product requirements.[2] When Esso discovered the Zelten and Raguba oilfields in the Liamco/Grace concessions, it honoured its commitment and signed an agreement for the construction of an 8000 b/d refinery at its export terminal at Marsa Brega. Half the production of this refinery would go to the local market where consumption was then around 4000 b/d, and half would go for export. At the time, Esso anticipated a 3 per cent return on its investment.

Brega Refinery. The Esso refinery was built in Belgium and shipped, more-or-less intact, to Marsa Brega in 1963.[3] It was ready to go on stream by mid-1964 but start-up was delayed for three years because of a dispute between Esso and the government on the pricing of products sold in Libya. Esso argued that it was not possible to operate on a commercially profitable basis with low local product prices because the cost of production at the Brega refinery was higher than that at larger European refineries. This was due to diseconomies of small-scale production, limitations on the type of crudes supplied in relation to product requirements and high transport costs to internal retail markets. Esso demanded local product prices based on refinery costs, transportation costs and return on investment, including cost of capital.

The government called in a British consultant firm, the Economist Intelligence Unit, for advice on product pricing. A compromise agreement with Esso on prices in the domestic market as well as on excise taxes on imported products was reached in early 1967. By this time, however, product demand within Libya had increased. The Brega refinery could meet only slightly more than half of Libya's requirements of gasoline, kerosene and diesel oil, although it was able to fulfil internal demand for fuel oil. Output was raised to 10,000 b/d, and there was talk of expansion.

While the Brega refinery was standing idle, the government sought agreements with European refiners, both state-owned and private, whereby these would provide products for the domestic Libyan market in exchange for guaranteed supplies of crude oil. In this effort it was unsuccessful, largely because it could not guarantee a regular supply of specific quantities of crude oil to any European refiner as the oil companies themselves handled all exports at this time.

Zawiyah Refinery. Even before the Brega refinery came on stream, the government had realized that it was inadequate for Libyan requirements. It decided there was a need for a second refinery with a capacity sufficiently large to meet rapidly increasing domestic demand and also to allow Libya to export products as well as crude. In mid-1969, it signed an agreement with Shell for the design, finance and construction of a 25,000 b/d refinery, capable of expansion to 40,000 b/d, at Zawiyah (Assawiyah), 31 miles west of Tripoli. Completion was scheduled for 1972. Under the terms of this agreement, Shell was to be reimbursed for the construction cost over a period of ten years, during which time it would manage the refinery. In addition, Shell was to build another plant at Zawiyah to produce 600 b/d of lubricating oil for consumption within Libya.

Negotiations with Shell over the details of the construction of this refinery were still in progress when the Revolutionary Command Council replaced the monarchy in September 1969. Shell was reluctant to proceed further with the proposed refinery as it saw its future in Libya as uncertain; to date, it had not found oil in its two concessions south of Tripoli or in its single concession in the western Ghadames Basin.

In the meantime, in order to meet increased internal product demand which was then more than 20,000 b/d, the Libyan National Oil Company concluded an agreement with Montedison whereby it would ship crude for processing to Montedison's Sincat refinery in Sicily, with products sent back to Libya. The agreement called for Sincat to process 30,000 b/d of Libyan crude, taking surplus naphtha in payment for a processing fee, and surplus fuel oil at cost. The agreement ran into trouble very early on, when the Italian authorities prevented Sincat from processing a large shipment of Libyan crude landed in Sicily because BP claimed that this had come from its Sarir field which the Libyan government had recently seized. It was only in March 1973 that an Italian court rejected BP's claim and the crude shipment was released to Sincat for processing.

In July 1970, the new government took over the import, distribution and sale of products on the local market, promising compensation to the oil companies whose businesses it had confiscated. The timing of this decision was undoubtedly influenced by an agreement it had recently concluded with the Algerian and Iraqi governments for a joint petroleum policy, as both of these countries already had nationalized their internal marketing facilities. The distribution assets which the government seized included terminals for imported refined products, storage facilities and 300 service stations, of which 104 belonged to Esso. Ownership of the others was divided almost equally between Shell and two Agip subsidiaries, Petrolibia and Asseil. The NOC set up a subsidiary, the Brega Petroleum Production Distribution Company, to manage the importation, transport, distribution and sale of imported products as well as of the government's share of products from the Brega refinery and any refineries built subsequently. It also commissioned a study of local marketing operations from a US consultant firm, Arthur D. Little.

Later that year, the government disclosed its failure to reach agreement with Shell over details of the construction of the Zawiyah refinery and its decision to proceed with the project on its own. By inviting bids from five engineering firms from Europe, the United States and Japan, the Revolutionary Command Council, despite its political rhetoric, continued the policy of the previous government of depending on Western companies for commercial projects. In June 1971, it signed a contract with Snam Progetti of Italy for the construction of a refinery at Zawiyah with an initial capacity of 60,000 b/d and a built-in capability of expansion. Completion was scheduled for August 1973. The refinery was to produce a range of products for the local market and for export. Shell, which was no longer involved in the refinery, agreed to draw up specifications for the lubricating oil plant, to be constructed by Mothercat under a contract signed in July 1972, with a mid-1973 completion date. Shell was to bear the cost of this plant in lieu of a signature bonus for new exploration acreage.

Having gone this far, the Revolutionary Command Council formulated a refinery programme on a grand scale. In November 1972, the minister of oil, Izzadin Mabrouk, announced a drive to make products, rather than crude, the most important Libyan export. The intention was to build refining capacity of 1 million b/d. With future product demand within Libya estimated at 70,000 b/d, this would leave a surplus of more than 900,000 b/d of products for export.[4] The minister mentioned increasing the capacity of the still incomplete Zawiyah refinery to 120,000 b/d if not 180,000 b/d and the construction of a new export refinery at Tobruk with a capacity of 130,000 or 150,000 b/d.[5] If future crude production reached 2 million b/d, as many predicted, this would mean that almost half of all Libyan crude would be exported as products, and the government did not foresee difficulties in selling these abroad. One of its plans was to 'oblige prospective customers of crude oil to acquire quantities of our refined products as well, thus assuring us a profitable outlet'.[6]

The refinery at Zawiyah was completed in September 1974 and even before it started commercial production, the government signed a contract with Snam Progetti to increase its capacity to 120,000 b/d. It announced its firm intention at this time to construct a 320,000 b/d export refinery at Tobruk, a 400,000 b/d export refinery at Zueitina and a 220,000 b/d export refinery at Misurata. Snam Progetti would build the Tobruk refinery and a Romanian firm the Misurata plant in accordance with an agreement made with the Romanian government. There was also talk of a second refinery at Brega with a capacity of 180,000 b/d and a Russian-built refinery of 600,000 b/d at Benghazi.

There were major structural changes in international products markets even as these ambitious plans were being made and refinery growth in Europe continued on its planned upward curve despite falling demand. In 1976, the Libyan government recognized the problem of overcapacity and announced a cutback in its refinery programme due to 'the world economic stagnation and Western dominance of the petrochemical and oil industries'.[7] It said that the Zawiyah refinery expansion would continue, but completion dates for the export refineries at Tobruk and Zueitina were put back from 1978 to 1980, and that for Misurata from 1980 to 1985.

Despite these cutbacks, the expansion in refinery capacity in Libya in the 1980s was very considerable. Between 1983 and 1987, domestic refining capacity in OPEC countries as a whole increased by 29.4 per cent; in Libya it increased by almost 200 per cent.[8] Libya lacked opportunities for the domestic investment of its income outside of the oil industry itself. There was little it could do to develop other industrial sectors due to a lack of raw materials and a small workforce, or to expand its agricultural sector due to lack of arable land and water. Investment in capital-intensive oil processing within the country seemed to many less risky at this time than foreign financial investments.

Ras Lanuf Refinery. In 1977, the government moved the site of the proposed export refinery at Tobruk to Ras Lanuf, partly as a result of increased hostile relations between Libya and Egypt and partly as a result of NOC's discovery of the large Messlah oilfield 12 miles north-west of the Sarir field. It decided to ship the output of this new field, as well as incremental production from the Sarir field, through the Sirtica pipeline to Ras Lanuf rather than send it to the terminal at Marsa Hariga, near Tobruk.

In October 1979, two ENI subsidiaries, Snam Progetti and Saipem, began the construction of the 220,000 b/d Ras Lanuf export refinery.[9] Completion was scheduled for 1981; the plant did not actually come on stream until 1984. By this time, construction of a petrochemical complex in Ras Lanuf was well underway and a new company, the Ras Lanuf Oil and Gas Processing Company (Rasco), was set up to manage both the refinery and the petrochemical complex.

In 1983, the government awarded a contract for the construction of a small refinery at Tobruk. The oil price collapse of the mid-1980s, however, put paid to the remainder of the government's ambitious domestic refinery plans, although a small topping plant was constructed later at the Sarir field. To a certain extent, the scaling down of plans for domestic refining capacity was a result of the government's decision to purchase refinery capacity in Europe.

Production Mix

Refining capacity within Libya, when the construction programme came to an end in 1988, was approximately 370,000 b/d. Appendix 8.1 gives details of the refineries which had been built and were running in Libya at this time.[10] It shows that the production of the largest refinery, Ras Lanuf, has varied considerably in recent years. Its throughput in 1988, for instance, was 156,000 b/d, rising to 190,000 b/d in 1989 and falling back to 160,000 b/d in 1990, when it was working at about 70 per cent of capacity. These production swings presumably reflect changes in the balance between Libyan crude and product exports. The other refineries were producing at capacity in the late 1980s except for Sarir, which only came on stream in 1988.

All these refineries, with the exception of the small Brega plant, produce an unusually high proportion of fuel oil. The emphasis on fuel oil appears to be the result of a decision based partly on the characteristics of the crudes Libya intended to process within the country. It was originally intended that the Zawiyah refinery would process Sarir crude, shipped by tanker from Tobruk, and would therefore have an output with a high proportion of fuel oil, as Sarir yielded in refining more fuel oil and less gasoline than other crudes of comparative gravity. Libya found it easier to process Sarir crude at home than to locate refineries abroad willing or even able to handle such a waxy crude.

The decision was also the result of the government's intention to use fuel oil for electricity generation. Libyan development plans, discussed in Chapter 10, focused on the construction of power stations and desalination plants powered by fuel oil. Demand for this product within the country in fact rose from 131.2 thousand tons in 1970 to 2112.9 thousand tons in 1985, an increase of 1510 per cent overall and an average annual increase of 20.3 per cent. Consumption increased particularly in the electricity sector where generating capacity rose by 1864 per cent between 1970 and 1986, an average annual increase of 20 per cent.[11]

There was also brisk demand abroad for Libyan heavy fuel oil, with its low sulphur content compared to most Middle East crudes, as a result of environmental legislation restricting the burning of high-sulphur fuel oil in power stations. Many regulations of this nature were made in the early 1970s; a maximum limit of 1.6 per cent sulphur content for fuel oil used in power stations in some areas of Germany, for instance, was set in 1971.

On the other hand, this decision left the industry without the capacity to produce an adequate supply of lighter products. Despite the fact that domestic demand for these (especially gasoline), rose 540 per cent between 1970 and 1985, there was no attempt to alter the product mix of the refineries in this period. Instead, the government sought to arrange contracts with European refineries and dealers for the exchange of Libyan fuel oil and naphtha for gasoline. Later, sanctions effectively came to prevent substantial changes aimed at creating a lighter output by forbidding the import of licences for upgrading technologies held by US firms.

Secure domestic and foreign markets for heavy fuel oil justified the high output of this product in Libyan refineries processing crudes such as Sarir. It is hard to argue in favour of a bias towards heavy fuel oil, however, in the 120,000 b/d Zawiyah refinery which in the late 1980s and early 1990s was processing Libya's best quality Es Sider crude.

Product Exports

Libyan exports of refined products were insignificant until the mid-1970s, when the Zawiyah refinery came on stream. There was another increase in the mid-1980s when the Ras Lanuf refinery began production, but exports have never reached the government target of 200,000 b/d, as Table 8.1 shows. To some extent, export volumes are restrained by the size of the ports at Ras Lanuf and Zawiyah.[12]

Table 8.1: Libyan Exports of Refined Products. Thousand Barrels Per Day

Year	Exports
1973	35.0
1974	27.8
1975	47.2
1976	53.0
1977	91.2
1978	99.2
1979	84.6
1980	72.5
1981	62.1
1982	78.0
1983	78.7
1984	76.8
1985	90.9
1986	101.9
1987	110.5
1988	90.0
1989	127.5
1990	150.0
1991	140.0
1992	170.0
1993	142.0
1994	130.0

Note: Figures after 1982 are estimates as of 15 July, 1995.

Source: *OPEC Annual Statistical Bulletins.*

The International Energy Agency publishes data on product imports by OECD countries which indicate that the majority of Libya's product exports, around 80 per cent on average, go to western Europe. Table 8.2 shows that naphtha, heavy fuel oil

Table 8.2: Libyan Product Exports to Europe. Main Product Type Share. Per Cent

Year	Product	Share	Year	Product	Share
1986	Naphtha	44	*1991*	Heavy Fuel Oil	31
	Heavy Fuel Oil	24		Naphtha	29
	Gas/Diesel Oil	23		Gas/Diesel Oil	28
1987	Naphtha	32	*1992*	Heavy Fuel Oil	37
	Gas/Diesel Oil	31		Naphtha	31
	Heavy Fuel Oil	26		Gas/Diesel Oil	21
1988	Gas/Diesel Oil	32	*1993*	Naphtha	33
	Naphtha	30		Heavy Fuel Oil	23
	Heavy Fuel Oil	23		Gas/Diesel Oil	25
1989	Heavy Fuel Oil	31	*1994*	Heavy Fuel Oil	36
	Naphtha	30		Naphtha	27
	Gas/Diesel Oil	26		Gas/Diesel Oil	17
1990	Heavy Fuel Oil	34			
	Gas/Diesel Oil	29			
	Naphtha	25			

Source: International Energy Agency, *Oil and Gas Information* 1986–8, 1987–9, 1992, 1994.

and gas/diesel oil have formed the bulk of Libyan product exports since 1986 to OECD countries, principally western Europe. The market for naphtha is well diversified and the naphtha export share which was 44 per cent in 1986 had decreased to around 33 per cent in 1993 and to 26 per cent in 1994. The share of heavy fuel oil exports, as might be expected due to European environmental regulations on sulphur content, increased from 24 per cent in 1986 to 36 per cent in 1994. Libyan gas/diesel oil sales, which are well balanced over several national markets, have varied considerably over the years, with no clear trend for either an increase or a decrease.

As can be seen in Table 8.3, the main European market for Libyan products is Italy, which takes large amounts of heavy fuel oil and considerable amounts of gas/diesel oil and naphtha. Second-tier importers are France, Germany, Spain and the Netherlands. France takes no fuel oil but considerable quantities of gas/diesel oil, naphtha and kerosene. Germany's imports are

Table 8.3: European Imports of Libyan Refined Products. Thousand Metric Tons

	1986	*1987*	*1988*	*1989*	*1990*
Austria	-	-	-	-	5
Belgium	176	-	134	74	308
Denmark	-	74	52	32	42
Finland	-	6	5	10	-
France	542	940	578	590	853
Germany	674	694	364	314	367
Greece	133	137	446	416	729
Italy	811	1163	1109	2230	2418
Netherlands	545	593	266	292	282
Norway	44	-	32	18	-
Portugal	-	115	42	50	-
Spain	408	431	321	441	512
Sweden	121	-	-	5	156
Switzerland	5	-	45	24	24
Turkey	-	-	-	-	-
UK	180	111	82	-	-
Total	3639	4264	3476	4496	5696

	1991	*1992*	*1993*	*1994*
Austria	-	-	-	-
Belgium	231	261	329	79
Denmark	28	15	15	139
Finland	-	-	4	-
France	908	911	716	686
Germany	458	661	357	119
Greece	341	166	124	260
Italy	2192	2562	984	1900
Netherlands	257	630	174	238
Norway	-	-	-	41
Portugal	3	2	-	2
Spain	639	874	320	378
Sweden	131	258	85	63
Switzerland	10	1	-	-
Turkey	20	-	72	111
UK	5	324	126	228
Total	5223	6250	3455	4244

Source: International Energy Agency, *Oil and Gas Information* 1986–8, 1987–9, 1992, 1994.

split almost evenly between naphtha and gas/diesel oil with, in some years, some imports of gasoline. Spain's imports are almost entirely naphtha. The Netherlands imports some of everything.

Expansion Plans

In late 1989, the government announced plans to expand its domestic refinery production. It decided to upgrade and increase the capacity of the Tobruk refinery and awarded a design contract to the Italian firm Compagnia Tecnica Internazionale Progetti (CTIP) and a construction contract to a Yugoslav company, Energoinvest. As of early 1986, there had been no reports of progress on this project, possibly because of the effects of the Yugoslav civil war on Energoinvest's ability to carry out the work. Originally, the upgrade was scheduled for completion in 1997.

In addition, the government decided to go ahead with the construction of a 20,000 b/d atmospheric distillation refinery at Sebha, processing crude from the Murzuq field when it came on stream, to supply fuel oil for a new power plant. In May 1989, it awarded an engineering design contract for this plant to CTIP; the design was completed in two years. In 1993, however, the government announced that construction of the Sebha refinery was on hold due to delays in the development of the Murzuq oilfield and a re-evaluation of the need for a power plant in the area. The beginning of Murzuq's development by the Repsol consortium in 1995 led to a reconsideration of the project.

In 1991, the government approved NOC's five-year plan which included increasing Libya's total refinery capacity, domestic and foreign, to approximately 1 million b/d by 1995. Within Libya, NOC's plans included increasing the capacity of the planned refinery at Sebha to 150,000 b/d and the construction of a new refinery at Hymed, on the north-west coast near the offshore Bouri field, with a capacity of 85,000 b/d. None of these projects had progressed beyond the drawing board by early 1996.

The five-year plan also included upgrading the existing refineries at Zawiyah and Ras Lanuf. The government admitted that the shortfall in light distillates production at these plants was a problem; the Tripoli area alone was importing up to 20,000 b/d of gasoline in the early 1990s. In April 1993, three engineering companies were invited to bid for a consultancy contract to upgrade Zawiyah. These included Libya's UK-based Teknica, Italy's Tecnimont and Canada's Monenco. A Croatian consortium, INA, was scheduled to do the upgrading of Ras Lanuf, which was to include the construction of a reformer for unleaded and premium gasoline and a hydrocracker to convert heavy fuel oil into jet fuel and diesel. This work was delayed by the effect of the Yugoslav war on INA. Sanctions regarding international trade with Libya, which are discussed in Chapter 10, undoubtedly are an important reason for the lack of progress in improving Libya's domestic refineries, as these prevent access to essential process licences for upgrading and limit the financial resources that the government has available for investment. The November 1993 UN Security Council ban on the supply of equipment for Libyan refineries apparently has had an impact not only on upgrading and expansion plans but also on day-to-day refinery operations.

Downstream Abroad

High oil prices in the late 1970s and early 1980s provided the Libyan government with a large surplus of oil income which enabled it to invest in downstream facilities in Europe. The wish to secure its own refinery outlets in European markets was not new; in July 1969, the Idris government had engaged in talks with West Germany regarding the construction, on a 50:50 joint-venture basis, of an oil refinery in Bavaria. A year later, the Qaddafi government announced that it was negotiating with West German companies concerning the building of oil refineries. Nothing came out of these talks.

By investing its petrodollars in downstream facilities abroad, Libya was following close on the heels of two other OPEC members.[13] One of these was Kuwait, which had a large surplus of petrodollars and which was particularly anxious to acquire refineries and distribution outlets in Europe in order to provide flexibility in the marketing of its crude production. In the early 1980s, before the discovery of light crudes in Kuwait, all Kuwaiti crudes were heavier than the average and contained a high percentage of sulphur. Both of these physical characteristics reduced their value in a market that sought light products which would burn clean. Kuwait also suffered a geographic disadvantage in that it was the Middle East oil producer farthest from Europe on the tanker routes and lacked pipelines to serve as an alternate means of delivery.

In addition, Kuwait's oil minister, Ali Khalifa al Sabah, wanted KOC to become a major international oil company. To this end, KOC started its overseas expansion in upstream activities; it engaged in several small joint-venture exploration deals in Africa and elsewhere and, in 1981, purchased Sante Fe International, a US-based exploration and drilling company. In the spring of 1983, after one year of negotiations, Kuwait purchased Gulf Oil's refining and distribution assets in Belgium, the Netherlands, Denmark, Sweden and Luxembourg. These included a 85,000 b/d refinery at Copenhagen and a 75,000 b/d refinery at Rotterdam, both later upgraded, as well as two lubrication oil plants, crude and product storage capacity and 1,500 service stations. Buying these made good commercial sense; Gulf had been one of the former shareholding partners in Kuwait and its European refineries were designed for Kuwaiti crudes. After leaving Kuwait, Gulf did not have sufficient crude of its own to feed its European refineries and lacked adequate financial resources to update its facilities.[14] A year later, Kuwait also purchased Gulf's assets in Italy. Subsequently it acquired Elf's service stations in Belgium, Hays Petroleum's service stations in the UK and Mobil Italiana, the latter including a 100,000 b/d refinery in Naples as well as a number of service stations. By 1990, it had 4800 service stations in Europe flying its Q8 banner.

In deciding to venture into European downstream operations, Libya had not only the example of Kuwait but also that of Venezuela. The Venezuelan national oil company, PdVSA, inaugurated its 'internationalization' programme in 1983 by entering into downstream operations with Veba Öl, a subsidiary of Veba. The two companies established a 50:50 joint venture, Ruhr Öl, which owned a 250,000 b/d refinery complex near Gelsenkirchen in West Germany. Two years later, the holdings of this joint venture increased to include shares in three other refineries as well as in a petrochemical complex, pipelines and terminals. At the end of 1986, Venezuela also acquired downstream facilities in the USA.

While it shared the concerns of Kuwait and Venezuela for the control of downstream outlets in Europe in order to protect their crude sales in times of weak markets, the good quality of Libyan crudes and the country's geographical proximity to Europe lessened the anxiety of the Libyan government on this point. It was acutely aware, however, in the early 1980s, of the probable total loss of the US market and felt keenly the need of cementing its European ties.

Not surprisingly, given the long-term close connection between the two countries, Libya's first purchases of foreign downstream assets were made in Italy. In 1983, the Libyan Arab Foreign Bank took part, with several other banks, in a loan to Tamoil Italia, a company formed by Roger Tamraz, a wealthy Egyptian-born financier then living in Lebanon who was rumoured to have numerous connections in the Arab world.[15] The loan was made to enable Tamoil to buy the refining and marketing assets of Amoco Italia. The many international major oil companies that were selling off their Italian downstream operations at this time included not only Amoco but also BP, Shell, Chevron, Texaco and Total. This was due partly to the need for large-scale investments to upgrade their Italian refineries to increase their light product yield and to meet tighter environmental regulations. In addition, profits were low due to the Italian government's setting of national product prices. In 1985, the Libyan Arab Foreign Bank purchased a 70 per cent controlling stake in Tamoil. The other shares were divided between Sasea, a group of Italian private sector investors, and Strand, an investment group controlled by Roger Tamraz.

Early in 1988, Libya established the Oil Investments International Company (Oilinvest), a state holding company registered in Curaçao in the Netherlands Antilles, to oversee 'all Libyan investments in oil, gas, petrochemicals and energy outside the country' and to develop an 'integrated downstream strategy' in the traditional markets for Libyan oil of north-west Europe and the Mediterranean.[16] Over the next 13 years, Oilinvest, through its many subsidiaries, acquired a majority interest in European refineries, distribution networks for refined products and petrochemicals and retail service centres. By the early 1990s, Libya, through Oilinvest, controlled 250,000 b/d of refining capacity and some 3000 service stations in several European countries. Appendix 8.2 gives details of the three European refineries it acquired. Unlike the Gulf refineries which Kuwait had purchased and which fitted the needs of their crudes, none of the Libyan purchases appear to have been based on a fit with Libyan oil such as, for instance, the high wax content of its Sarir and other crudes.

Italy. Oilinvest increased its control over Tamoil to 85 per cent in 1989 and to 100 per cent by 1991. The original Tamoil assets included 800 Italian service stations and the former Amoco refinery located in Cremona in the Po valley. This had been built in 1954 with an original capacity of 60,000 b/d, expanded in the early 1970s to 100,000 b/d. In 1987, Tamoil purchased the former Texaco Italy marketing network containing 770 service stations. In the next few years, it enlarged its retail distribution network by purchasing 40 Fina service stations in Sardinia and a number of service stations located in southern Italy from the Cameli group. It also bought several retail outlets in Sicily from independent local organizations. By 1993, Tamoil controlled 1939 service stations in Italy and held a 5.5 per cent share of the Italian gasoline retail market. Tamoil diversified under Oilinvest's management. In 1986 and 1987, it acquired a 50 per cent, later increased to 87 per cent, share in F.A. Petroli, a company operating in the wholesale product trade in northern Italy with a bulk terminal plant near Milan connected by pipeline to the Cremona refinery. In 1988, it acquired a 75 per cent interest, later increased to 100 per cent, in Vulcan Oil, a small, Milan-based sales company dealing in the wholesale gas oil and fuel oil markets. In 1989, it took a 50 per cent, later increased to 75 per cent, and in 1994 to 100 per cent, shareholding in A. Bortolotti, a petroleum products marketing company with sales in seven provinces and in 1990, purchased a half share of Sirm, a Rome-based company involved in the design, construction and maintenance of service stations. In 1990, Tamoil restructured, consolidating Vulcan Oil, Petroli and Bortolotti in a newly-formed company, Tamoil Petroli Italiana, jointly owned by Tamoil and Oilinvest.

In the early 1990s, Tamoil was serving as an important outlet for Libyan crude oil. In 1992, for instance, it sold 60,000 b/d of Libyan crudes, mostly Brega and Es Sider, on the open market and imported some 69,000 b/d for its Cremona refinery. In addition, it imported petroleum products from Libya, mainly gasoil and fuel oil. In the early years, the Cremona refinery mainly processed Es Sider crude although it also took small quantities of the Brega, Sirtica and Amna blends. In 1990, a dewaxing unit was installed allowing it to process Sarir crude which, in 1994, accounted for more than 22 per cent of its crude intake. Only 14 per cent of the feedstock in this year consisted of non-Libyan crude oils, such as Ural and Siberian Light.[17]

The Cremona refinery's most important product is fuel oil. Its product line in the mid-1990s varied between 31 to 35 per cent heavy distillates, 45 to 46 per cent medium distillates and 19 to 24 per cent light distillates. Direct retail sales in Italy accounted for 20 to 27 per cent of its output, and exports of gasoline, gasoil, jet fuel, naphtha and lubricants, primarily to Switzerland, between 3 and 10 per cent. Cremona is the main supplier of low-sulphur fuel oil for power generation in northern Italy. Two generating plants of the state-owned Italian utility Enel, at Fermide and Ostiglia in the Po valley, are directly connected to the refinery by pipeline. Most Italian refineries can only meet the Italian regulation that fuel oil cannot exceed a 0.25 per cent sulphur content by processing very low sulphur crudes, such as Es Sider.

Germany. In 1988, Oilinvest acquired a 30 per cent equity holding in the 80,000 b/d Holborn refinery in Hamburg from the Coastal Corporation of Houston which had acquired the refinery from Esso the year before. Coastal's deal with Oilinvest was struck in settlement of a $45 million financial commitment which Coastal had undertaken in 1980, and subsequently failed to fulfil, for exploration and development of a concession in Libya. Three years later, Oilinvest exercised its right to accelerate the conversion of shares from Coastal in its favour, and increased its holding to a majority share of 67 per cent in the newly formed Holborn Investment Company. The refinery itself was badly in need of upgrading, and Oilinvest later spent quite a sum on this work.

The yield of the Holborn refinery is different from that of Cremona. It averages only 15 per cent heavy distillates, with the rest of its product line light and medium distillates – 25 per cent diesel, 25 per cent light heating oil, 20 per cent gasoline and 6 per cent jet fuel. Its sales are handled by Holland Marketing of Rotterdam, a subsidiary of Hamburg Eggest Mineraloelhandels (HEM) with most products sold in Germany. A few are exported to the Netherlands and a considerable amount of the heavy fuel it produces is exported to Scandinavia. NOC supplies the refinery with 75,000 b/d of crude oil on average, principally Es Sider, on a netback basis and Holborn provides NOC with access to storage for 4.8 million barrels of crude and 1.65 million barrels of refined products. It owns the oil pipeline from Wilhelmshaven to Hamburg and in 1989, purchased a 15 per cent share in the Wilhelmshaven-to-Cologne pipeline group Nord-West Oelleitung (NWO), giving it rights to terminal loading and tank storage facilities at the North Sea port of Wilhelmshaven.

In 1991, Oilinvest took an 80 per cent stake, later raised to 90 per cent, in HEM, the German product distribution company, changing its name to the Holborn European Marketing company.[18] At the time, HEM owned about 400 service stations which depended on the Holborn refinery for supplies. By 1994, HEM controlled a network of some 450 retail outlets throughout Germany and had announced plans to expand the number of its outlets in eastern Germany from 75 to 120.

Switzerland. In 1990, Oilinvest, through Tamoil, purchased a 65 per cent stake, later expanded to 82.5 per cent and subsequently to 100 per cent, of Gatoil, a Swiss company then under court-appointed trustee management as a result of the debts of its owner, Khalil Ghattas. The assets of Gatoil, renamed Tamoil SA, included 329 service stations, 14 terminals and a refinery at Collombey, in the south-west of the country, near Lake Geneva.[19] Collombey's theoretical capacity was 70,000 b/d but it was only capable of producing 25,000 b/d when it was shut down in 1990. Once again, Oilinvest had selected a dilapidated plant and Tamoil had to make a considerable investment in its restoration and upgrading, including the addition of isomerization and desulphurization units. In the early 1990s, the refinery was processing mainly Brega blends which were landed at Genoa and shipped directly to Collombey via the Italian Snam pipeline. In 1991, Tamoil SA began supplying oil products from Collombey to the 280 service stations of Migrol, a subsidiary of Switzerland's second-largest retail distributor.[20]

Malta. Oilinvest acquired Chempetrol Overseas Ltd, a distributor of petrochemicals, in 1988. Its principal activity is the marketing and sale of methanol, which Libya produces, to Italy, Spain, the Netherlands, Portugal, France and Brazil.

Spain. Oilinvest established Oilinvest España, based in Madrid, in 1991, with the intention of entering into the Spanish retail service station market. At the time, it hoped to take a 75 per cent interest in the retail operations of Estaciones de Servicio Guipuzcoanas (Eserguil), a small marketing and distribution network with about 30 gasoline outlets in the Basque region, near Barcelona. Agreements with Spain's Repsol and Enagas for cooperation in upstream and downstream ventures apparently included an Eserguil shareholding purchase, but the Spanish government blocked the sale. In 1992, Oilinvest launched a retail marketing programme in Spain on its own, with plans to open 20 gas stations annually.

France. Oilinvest established Oil Energy France in 1991, with the intention of acquiring Theveni & Ducrot, a small French independent oil distributor with a 14,000 b/d distribution system and 360 stations in eastern France. Although the acquisition did not succeed at this time, Tamoil has managed to sell products in France through Oil Energy France.

Eastern Europe. Libya began exporting crude and products to eastern Europe through barter trade in the 1970s, and in 1991 about half of the 200,000 b/d of the crude sent to this area was still on barter terms. In the early years, barter arrangements involved trade in commodities; later they involved payment for engineering and construction services rendered in Libya by eastern European companies. In the late 1980s, Libya decided on a strategy of asset acquisition in the region. Overtures which offered assurance of long-term crude supply in return for a minority stake in refineries seemed to interest eastern European governments more than bids from major oil companies for a controlling interest in refineries.

In 1991, Oilinvest established a product distribution company, Tamoil Hungaria, as a joint venture between Oilinvest, with 75 per cent of the shares, and Hungary's Mineralimpex and Austria's Mineralkonter. The new company embarked on a programme of expanding retail services supplied with Libyan crude and products from Oilinvest's refineries in Germany, Italy and Switzerland. In 1992, Oilinvest purchased two retail outlets in Slovakia and took an 80 per cent share in Temp, a Czech Republic service station chain. In 1994, the newly formed Tamoil Praha announced plans to acquire 10 per cent of the retail Czech market and to build new retail outlets over a period of 12 years. At the same time, Oilinvest announced plans to further increase its outlets in eastern Germany and the Slovak Republic. Negotiations for a stake in a Yugoslav petrochemical complex, however, were abandoned as a result of the war in Yugoslavia.

Greece. Libya had arrangements for processing crude at the 100,000 b/d Motor Oil Hellas refinery at Corinth in Greece on an intermittent basis from the early 1980s. In 1991, Oilinvest launched an attempt to buy a minority stake in this refinery; negotiations had not come to any conclusion by early 1996.

Egypt. In 1992, Tamoil took an 80 per cent interest in a company, Stamoil, a joint venture with the Egyptian General Petroleum Corporation and several private Egyptian investors. In addition to acquiring 17 service stations in Egypt, Stamoil is involved in a proposed products pipeline from Libya to Alexandria.

Cutbacks in Libyan Investment in Europe

In the early 1990s, the Libyan government was planning to expand Oilinvest's downstream oil refining and marketing systems in Europe to handle 400,000 to 450,000 b/d. The chairman of Oilinvest, Mohammed Abdul Jawad, spelled out the continued benefits of downstream investment in Europe in an interview published in the *Middle East Economic Survey* in May

1990.[21] He cited the security of market outlets, the value-added factor in product sales, diversification of profit centres and increased expertise in oil pricing and marketing. At this date, Libya was channelling an average of 213,000 b/d of crude oil through its integrated downstream refining and marketing outlets in Europe and was one of the bidders for a minority holding of the state-owned oil company OMV that the Austrian government was planning to sell.

The threat of increased sanctions against Libya, however, forced a change in plans. In October 1993, Libya reduced its share of Oilinvest to 45 per cent in favour of some of its European partners in order to protect the company and its subsidiaries from the effects of the impending UN sanctions. Control of Oilinvest was shared with the family-owned Armani Group along with seven other European private business groups. A great deal of portfolio realignment followed at the time and subsequently, involving the various subsidiaries of Tamoil, Oilinvest, and Holborn European Marketing. By late 1994, however, some of the changes seemed to be reversing. Early in 1995, *Petroleum Intelligence Weekly* reported that Oilinvest's trading division, Tamoil Trading, which had effectively ceased functioning after the 1993 shake-up, had re-emerged as a more active player in both crude and product markets.[22]

Notes

1. 'Libya Envisages Downstream Integration for National Oil Corporation', *Middle East Economic Survey Supplement*, IX/42, 19 August, 1966.
2. See Kingdom of Libya, Ministry of Petroleum Affairs, *Libyan Oil 1954–1967* (1968), pp. 71–2.
3. *An-Nahar Arab Report & Memo*, Beirut, 4 September, 1978.
4. In fact, domestic demand for refined products grew by leaps and bounds during the 1970s and early 1980s and peaked in 1982 at 118,800 b/d; subsequently it declined to a low of 94,000 b/d in 1986. By early 1993, it had risen again to 110,000 b/d. *OPEC Annual Statistical Bulletin*, 1993.
5. *Middle East Economic Digest*, 20 November, 1972, p. 1301.
6. *OPEC Bulletin*, IX/38, 18 September, 1978, p. 6.
7. *Middle East Economic Digest*, 20/3, 16 January, 1976, pp. 18–19.
8. Bob Williams, 'OPEC Ventures Downstream: Industry Threat or Stability Aid?', *Oil & Gas Journal*, 16 May, 1988, p. 15.

9. Specifications called for an atmospheric distillation unit, a naphtha hydrosulphurization unit, a catalytic reformer, and an LPG recovery and treatment unit.
10. The government sometimes lists the processing plants which remove volatile fractions at the Nafoora, Amal, Intissar, Zueitina, Dahra fields, with capacity varying between 1800 b/d and 3000 b/d, as refineries.
11. See Committee for Middle East Trade, *Libya: The Five Year Development Plan 1981–85* (1982), p. 28; *Middle East Economic Survey*, 29/3, 28 October, 1985 p. 7; 31/26, 4 April, 1988, p. D9; 31/42, 25 July, 1988, p. 2.
12. Maximum loading of gas oil ex Ras-Lanuf is 25,000 tonnes; liftings ex-Zawiyah range between 18,000 and 20,000 tonnes.
13. Other oil-producing countries that invested later in downstream operations outside their borders included Saudi Arabia, which preferred joint ventures to complete ownership, and Mexico.
14. Prior to finalizing its deal with KOC, Gulf had disposed of a large part of its other downstream operations in Europe and Asia and was in the process of reorganizing as a primarily domestic US company.
15. In late 1995, Roger Tamraz was identified as having seriously considered investing £400 million to save the Maxwell Group in the UK from bankruptcy. In the event, he failed to do so.
16. NOC, Libya's national oil company, the Libyan Arab Foreign Bank, the Libyan Central Bank and the Libyan Arab Foreign Investment Company were the initial shareholders. See *Middle East Economic Survey*, 31/30, 9 May, 1988, pp. A3–A4 and 31/48, 12 September, 1988, p. A3.
17. For information on Tamoil, see Tamoil Italia S.p.a. *Financial Statements and Reports*, and *Middle East Economic Survey*, 1988–1995.
18. Oilinvest purchased HEM from the Hamburg-based trader Marimpex, which sometimes handles Libyan crude sales.
19. The only other refinery in Switzerland at this time was the Shell 60,000 b/d Cressier refinery in the north, near Berne. Domestic refining accounted for 35 per cent of Swiss consumption.
20. Migrol and Sasea, the Italian investment group originally connected with Tamoil, were involved in the original purchase of Gatoil.
21. *Middle East Economic Survey*, 33/32, 14 May, 1990, pp. A4–A5.
22. *Petroleum Intelligence Weekly*, 20 February, 1995, p. 5.

APPENDIX 8.1

LIBYAN REFINERIES IN LIBYA

RAS LANUF REFINERY

Location: Ras Lanuf, Gulf of Sirte

Start-up: 1984

Capacity: 200,000 b/d

1988 Utilization: 69.3 per cent

Employees: 2401

Crude Supply: Mesla & Sarir: 38.4°API, 0.1–0.17 swt

1988 Production

Product	*Actual Percentage*	*Design Percentage*
LPG	1.0	0.9
Gasoline	0	0
Naphtha	20.3	16.4
Kerosene/Jet Fuel	5.3	5.0
Gasoil	23.3	24.1
Fuel Oil	44.4	53.6
Other	0.1	0
Plant Fuel/Losses	5.4	0
Total	99.8	100.0

ZAWIYAH REFINERY

Location: Zawiyah, on Mediterranean coast west of Tripoli

Start-up: 1975

Capacity: 120,000 b/d

1988 Utilization: 100 per cent

Employees: 1396

Crude Supply: Es Sider (75 per cent) 36.9°API, 0.4 per cent swt
Hamada (25 per cent) 39.1°API, .05 per cent swt

1988 Production

Product	*Actual Percentage*	*Design Percentage*
LPG	0.3	0.4
Gasoline	9.4	9.5
Naphtha	6.6	7.8
Kerosene/Jet Fuel	11.7	13.8
Gasoil	18.9	17.9
Fuel Oil	45.8	47.0
Other	2.8	0
Plant Fuel/Losses	4.5	3.6
Total	100.0	100.0

SARIR REFINERY

Location: Inland at Sarir oilfield

Start-up: 1988

Capacity: 10,000 b/d

1988 Utilization: 26.4 per cent

Employees: 303

Crude Supply: Sarir 37.4° API, 0.14 per cent swt

1988 Production

Product	*Actual Percentage*	*Design Percentage*
LPG	0	0.1
Gasoline	0	10.0
Naphtha	21.6	5.3
Kerosene/Jet Fuel	1.7	2.8
Gasoil	27.6	30.2
Fuel Oil	49.1	50.5
Other	0	0
Plant Fuel/Losses	0	1.1
Total	100.0	100.0

TOBRUK REFINERY

Location: Tobruk, on Mediterranean Sea

Start-up: 1986

Capacity: 20,000 b/d

1988 Utilization: 100 per cent

Employees: 221

Crude Supply: Sarir 36.6° API, 0.24 per cent swt

1988 Production

Product	*Actual Percentage*	*Design Percentage*
LPG	0.7	0.9
Gasoline	0	14.4
Naphtha	16.1	0
Kerosene/Jet Fuel	8.1	9.4
Gasoil	20.9	25.5
Fuel Oil	50.5	49.0
Other	1.2	0.8
Plant Fuel/Losses	2.4	0
Total	99.9	100.0

BREGA REFINERY

Location: Brega, on Gulf of Sirte

Start-up: 1970

Capacity: 10,000 b/d

1988 Utilization: 100 per cent

Employees: 35

Crude Supply: Brega - 41°API

1988 Production

Product	*Actual Percentage*	*Design Percentage*
	0	0
Gasoline	8.2	18.5
Naphtha	23.7	1.7
Kerosene/Jet Fuel	0.1	11.4
Gasoil	28.2	24.9
Fuel Oil	35.2	38.1
Other	0	0
Plant Fuel/Losses	3.9	5.4
Total	99.3	100.0

Source: Organization of Arab Petroleum Exporting Countries, *Prospects of Arab Petroleum Refining Industry: Handbook of Arab Oil Refineries* (1990), pp. 94–108.

APPENDIX 8.2

LIBYAN EUROPEAN REFINERIES

CREMONA REFINERY

Location: Cremona, Italy

Total Crude Distillation Capacity (b/d):	95,000
Charge Capacity (b/cd):	
Vacuum Distillation	
Thermal Operations	33,000
Catalytic Cracking	——
Catalytic Reforming	13,400
Catalytic Hydro-cracking	6,000
Catalytic Hydrorefining	11,500
Catalytic Hydrotreating	1,800
Production Capacity (b/cd):	
Alkylation Polymerization	——
Aromatics/Isomerization	4,400
Lubes	——
Asphalt	——
Hydrogen (mcf/d)	——
Coke (t/d)	——

HOLBORN REFINERY

Location: Hamburg, Germany

Total Crude Distillation Capacity (b/d): 150,000

Charge Capacity (b/cd):	
Vacuum Distillation	53,900
Thermal Operations	28,000
	23,000
	19,900
Catalytic Cracking	——
Catalytic Reforming	32,000
Catalytic Hydro-cracking	——
Catalytic Hydrorefining	73,000
Catalytic Hydrotreating	47,600

Production Capacity (b/cd):	
Alkylation Polymerization	——
Aromatics/Isomerization	——
Lubes	——
Asphalt	10,000
Hydrogen (mcf/d)	——
Coke (t/d)	1,200

COLLOMBEY REFINERY

Location: Collombey, Switzerland

Total Crude Distillation Capacity (b/d):	60,000
Charge Capacity (b/cd):	
Vacuum Distillation	24,000
Thermal Operations	11,000
	9,000
Catalytic Cracking	——
Catalytic Reforming	10,000
Catalytic Hydro-cracking	——
Catalytic Hydrorefining	——
Catalytic Hydrotreating	16,700
Production Capacity (b/cd):	
Alkylation Polymerization	——
Aromatics/Isomerization	3,600
Lubes	——
Asphalt	5,200
Hydrogen (mcf/d)	——
Coke (t/d)	——

Source: Petroleum Economist & Kennet Oil Logistics, *Oil Logistics Guide to Northern Europe and the Mediterranean 1995* (1995), pp. 146, 157, 171.

9 NATURAL GAS

A number of foreign companies began to show interest in Libyan natural gas reserves in the early 1990s. This was a new development. Of all the original concession holders in the early days of the Libyan oil industry, only Esso had turned its mind to the export potential of Libyan natural gas. For thirty years, no other foreign oil company, with the exception of Agip, showed any interest in becoming involved in the development of Libya's proved and potential gas reserves. There were a number of reasons for this. They considered such reserves to be insignificant compared to those of neighbouring Algeria and noted that production for many years was mainly from associated fields. In addition, they were unimpressed by Libya's single, somewhat antiquated LNG plant and the absence of a gas export pipeline. And finally, they noted the poor track record of Libya's LNG exports.

This jaundiced perception began to change following the discovery of sizeable non-associated gas fields in Libya, especially as this occurred at a time when unease was growing over the dependability of Algerian gas exports. The interest of foreign companies in developing these new fields was encouraged by the Libyan government.

Reserves

Estimates of Libya's gas reserves are given in Table 9.1. It is impossible to tell how this reserve basis is allocated between associated and non-associated fields as the Libyan government has not issued any field reserve statistics since the early 1970s.

There is some information available about Libya's known non-associated gas fields, shown in Table 9.2, of which three were in production in 1995. One of these, Hateiba, was put into production by Esso in the 1970s and there are no current figures as to its remaining reserves. The two others, Assumud and Sahl, began producing in the early 1990s; Assumud was then reported to have reserves of 3 trillion cubic feet (tcf), but there has been no official report on Sahl's reserves.

The non-associated fields that had been discovered but not

Table 9.1: Natural Gas Reserves. Billion Cubic Metres

Year	*Reserves*	*Year*	*Reserves*
1950	0	1984	562
1960	30	1985	570
1970	735	1986	626
1975	750	1987	728
1976	745	1988	727
1977	730	1989	827
1978	730	1990	1218
1979	685	1991	1208
1980	680	1992	1309
1981	675	1993	1299
1982	611	1994	1289
1983	566	1995	1310

Source: *Cedigaz*.

Table 9.2: Non-associated Gas Fields.

Fields in Production

Field	*Reserves Trillion Cubic Feet*	*Production Million Cubic Feet Per Day*	*Start of Production*
Hateiba	n.a.	240	1980s
Sahl	n.a.	148	1991
Assumud	3	75	1993

Fields Discovered and Undeveloped

Èield	*Estimated Reserves Trillion Cubic Feet*	*Estimated Production Million Cubic Feet Per Day*
Attahadi	9–10	250–350
NC-41	n.a. ('huge')	700
Al Wafaa	7	n.a.

developed by 1995 have considerably larger reserves than any of the fields in production. Attahadi, located in the central Sirte Basin, has reserves rumoured to be in the vicinity of 9–10 tcf. It was discovered by Sirte Oil, a subsidiary of the Libyan National

Oil Company, sometime in the 1980s but the government did not authorize its development until 1991 when plans were drawn up for initial production of between 250 and 350 million cubic feet per day (mcf/d), with the output delivered to the gas terminal at Brega through a new 50-mile, 36-inch pipeline. Attahadi's development was further delayed, however, apparently by government financial restraints and possibly by the effect of US and UN sanctions on the importation of production equipment and process licences.

The reserves of the non-associated offshore field NC-41, discovered by Agip in the mid-1980s, have never been reported but are said to be 'huge'. Ten gas structures have been identified within this block which lies offshore Tripoli, adjacent to the producing Bouri oilfield. The development timetable of NC-41 is vague, but a joint Agip and NOC feasibility study, completed in 1992, concluded that production of 700–850 mcf/d (8–10 billion cubic metres per annum), including production from nearby NC-151, was possible.[1] Agip, the operator of both Bouri and NC-41, has a 30 per cent oil interest, and the Libyan National Oil Company a 70 per cent oil interest in NC-41. Up to the mid-1990s, Agip had not finalized terms for a gas production sharing agreement on NC-41. Development was also hampered by the effect of sanctions, by political problems in Italy's public companies, and by budgetary constraints.

The NC-41 feasibility study recommended processing the produced gas at an onshore gas gathering and purification station rather than offshore. It called for the export of the processed gas via a new 350-mile undersea pipeline to Sicily where it would then enter the Italian national gas grid system. Realization of the full production potential of NC-41 may depend on Agip's larger plans for a Mediterranean pipeline gas grid including exportable production from all of North Africa. Elements of this scheme, to which Libyan production could be linked in theory, include the Gazoduc Maghreb Europe (GME) line from Algeria under construction in the mid-1990s, a proposed Egypt to Turkey pipeline via Israel and an enlarged TransMed line from Algeria to Italy via Tunisia, as well as a direct Libyan link to Italy. In late 1994, Agip was holding discussions with the Algerian, Tunisian and Libyan governments regarding a pipeline link.

The 1991 discovery of a second large onshore non-associated gas field, which aroused the interest of private companies because they will be needed to develop it, was initially shrouded in secrecy. What became clear – although not at the time of the announcement of the discovery by Sirte Oil – is that there is a large, non-associated gas/condensate field in the Ghadames Basin, near the Algerian border. This Al Wafaa field is believed to be a north-east extension of the Algerian Alrar field. Its reserves, originally given as 4 tcf of gas and 250 million barrels of condensate, have been increased to at least 7 tcf of gas, if not more.

The Al Wafaa field lies close to blocks for which Agip is, and Total was, listed as rightholders, and both of these companies were said to be interested in developing it. The Libyan government, however, was reluctant to assign Al Wafaa to either company. It felt that Agip had so many projects going and planned elsewhere in North Africa as well as in Libya that speedy development of Al Wafaa might not be its central concern. It holds the same sort of reservations regarding the priority that Total would give to this single project. It is said to be hoping to arrange a consortium for the development of the field.

The wide geographical distribution of Libya's non-associated gas fields poses difficulties for their development. The Al Wafaa field lies some 550 miles from the Sahl gas processing complex in the Sirte Basin and the gas terminal at Brega, and more than 300 miles from the Tripoli area intended as the processing centre for offshore NC-41 and NC-151 gas. The most economical solution for the export of Al Wafaa gas would be to send production to the treatment plant and pipeline terminus at the nearby Algerian Alrar field. Built in the mid-1980s by a Soviet company, the Alrar pipeline is connected, by stages, to Hassi R'Mel, from whence all Algerian export routes radiate. Several foreign oil companies which are active in Libya, including Total, Agip, Repsol and Lasmo, have taken positions in various Algerian licensing rounds and these might be well placed to develop a Libyan/Algerian gas production link. But any such arrangement would have to be agreed between the governments of Libya and Algeria and co-operative agreements between the two governments have been few and far between. Another solution would be to combine Al Wafaa output with that of offshore NC-41 and NC-151 in the proposed gas treatment plant near Tripoli for eventual shipment in the proposed Agip pipeline to Italy. In the mid-1990s, neither of these solutions appeared feasible in the short term.

The discovery of Al Wafaa intensified the focus of both oil and gas exploration in the 1990s on the western Ghadames and Murzuq Basins. Geologists have long insisted that Algerian gas and oil formations, such as the Illizi (former Polignac) Basin, extend into Libya where there are similar structural features. For years Esso, Shell, Amerada, Gulf, Total and Agip have searched in vain along the border for oil; presumably if they found gas in the early days, they ignored it. Most companies, when they made major discoveries in the central Sirte Basin, relinquished their border area concessions to concentrate on developing the infrastructure for exporting their Sirte Basin oil production.

Interest in the western area increased in the 1980s after NOC had discovered oilfields in the Ghadames Basin, and east European companies had found extensive oil reserves in the Murzuq Basin. Following its discovery of Al Wafaa, Sirte Oil extended its search for gas to the south, into the Tadrat Acacus region and Murzuq Basin.

There is apparently considerable non-associated gas in the 7 November Block which straddles the maritime boundary between Libya and Tunisia.[2] By the mid-1990s, little had been done to verify the gas reserves in this area or to examine the feasibility of extracting them. Libya and Tunisia had formed a Joint Oil Company to undertake exploration of this block in 1988, but shortage of funds limited its activity to the identification of interesting structures. In the early 1990s, the Joint Oil Company was reported to be seeking to form a joint venture, and a number of international oil companies were said to have expressed interest.

Gross Production

Esso began producing gas for pipeline shipment to its Brega gas treatment plant in the early 1970s from its two associated fields located in the central Sirte Basin, Zelten and Raguba. As most of the gas produced until 1990 came from associated fields, production amounts were largely determined by crude oil output. As Table 9.3 shows, gross production peaked in 1979, when the operations were still under Esso's control, at 23.4 bcm per annum. It then fell to an average level of 12 to 12.5 bcm per annum throughout most of the 1980s, except in 1986 when it was somewhat higher. In the early 1990s it increased, sometimes reaching 15 bcm per annum.

Between 1975 and 1985, about half of gross production was reinjected, with the share beginning to decline very slightly after 1983. In 1986, the proportion dropped abruptly as a result of increased gross production, presumably after the start-up of the over-saturated offshore Bouri field. Subsequently, reinjection

Table 9.3: Natural Gas Production. Billion Cubic Metres Per Annum

Year	*Gross Production*	*Reinjected/Other Industry Use*	*Flared*
1970	19.4	4.0	15.1
1975	13.9	6.3	3.0
1976	18.0	9.0	4.2
1977	20.0	10.6	4.3
1978	21.2	10.8	5.3
1979	23.4	11.9	4.7
1980	20.4	10.6	4.6
1981	12.7	6.6	2.3
1982	12.2	6.3	2.5
1983	13.3	6.5	3.3
1984	13.5	6.3	3.5
1985	13.5	6.0	2.9
1986	15.0	6.0	3.4
1987	12.0	4.8	2.2
1988	12.5	4.4	2.6
1989	13.5	4.4	2.3
1990	15.0	5.0	3.8
1991	15.8	5.6	3.7
1992	13.4	4.6	2.0
1993	12.4	2.3	1.7
1994	12.5	4.3	1.8

Source: *Cedigaz*.

fell to an average of about 30 per cent.[3] Flaring, except in 1986, also fell from a 1970 high of 78 per cent to 22 per cent in 1980 and to 14 per cent in 1994. The amount of production which has been marketed either within the country or as exports rose to more than half of gross production after 1992.

Sirte Oil, which took over the former Esso holdings, including the Brega LNG plant, is responsible for Libyan gas production and export. In the early 1990s, it reported producing some 723 mcf/d from fields which lie within a few miles of each other in the central Sirte Basin. Slightly more than one-third of this output came from the two former Esso associated fields, Zelten (200 mcf/d) and Raguba (60 mcf/d). The former Esso non-associated field, Hateiba, contributed 240 mcf/d. In 1991, Sirte Oil brought on stream first one non-associated field, Sahl, at 148 mcf/d and then, in 1993, another non-associated field, Assumud, at 75 mcf/d.

The hub of Sirte Oil's operations is the Sahl field, which it discovered in 1983. Sahl has 15 operating wells producing 150 mcf/d of natural gas and 4000 b/d of condensates; its gas treatment and processing plant began operations in 1991. As Figure 9.1 shows, Sirte pipes gas from the Sahl complex to the coastal terminal of Brega where some of it goes to petrochemical plants, some enters the LNG plant for export and the rest is sent westward in a 416-mile coastal pipeline to Khoms, near Tripoli. Construction of this coastal line, with a design capacity of 420 mcf/d, was begun by a Soviet company and eventually completed in 1989 by a Libyan company. It has feeder lines to Libya's main development projects, the petrochemical plants at Ras Lanuf, the fertilizer plants at Brega and the iron and steel plant at Misurata. It also serves several power stations and cement factories located along its route. Gas use in power stations only began after the completion of this pipeline; the government now plans to substitute gas for fuel oil in many existing power stations and to build new base-load, gas turbine power plants.

In 1990, a company from the former Yugoslavia, Razvoj i Inzinjering, was awarded a design contract to extend the line 149 miles along the coast through Tripoli to Bu Kammash, a town close to the Tunisian border. There are plans to extend it eastward to Benghazi. In addition, there are plans for the

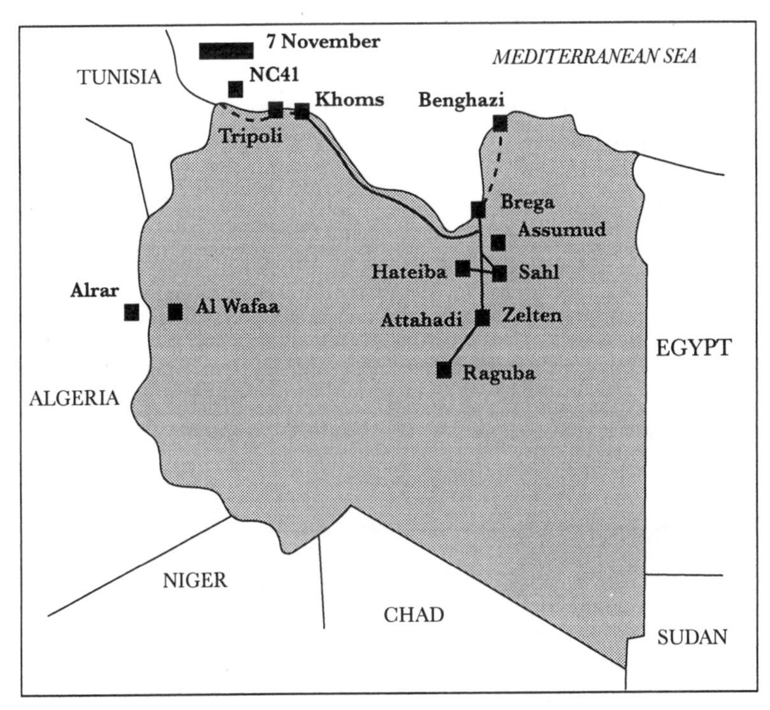

Figure 9.1: Libyan Gas Fields and Pipelines

construction of facilities for the use of associated gas at several oilfields. The Bu Attifel field, for instance, is to have a gas processing plant able to handle 8.49 mcm/d of gas and 20,000 b/d of condensate.[4] Most of the gas processed at this plant will be reinjected into the Bu Attifel reservoirs, but some will be piped to fields operated by the Zueitina Oil Company and some will be piped to Sahl. Gas from the Nafoora oil field will be piped to Bu Attifel after the completion there of a gas collection plant; a contract for this plant and for compressors was awarded in January 1994 to a German company, MAN.[5] A contract was signed in the early 1990s, also with MAN, for a gas treatment plant and compressors at the Sarir field. This gas will be piped some 90 miles to a projected power station in the Great Man-Made River project.[6] A contract has also been signed with the Italian company Bonatti for a gas treatment plant at the non-associated Assumud field.

Marketed Production

Table 9.4, which lists the percentage of gross production consumed within Libya and the percentage exported as LNG, shows a definite increase in the use of natural gas within the country since 1979. Libya's population is located almost entirely in coastal regions and very little, if any, gas is marketed elsewhere. It is primarily used in the industrial sector, in electricity generation and in Libya's petrochemical and fertilizer industries. Very little gas goes to the residential sector of the economy and there is no pipeline connection to the main cities of Tripoli and Benghazi. The proportion of marketed production exported as LNG has varied in accordance with pricing and marketing conditions.

Table 9.4: Marketed Production. Percentage of Gross Production

Year	*Exported as LNG* *Per Cent*	*Consumed within Libya* *Per Cent*
1975	23.1	10.9
1976	20.0	7.0
1977	17.9	7.1
1978	16.1	7.9
1979	12.9	16.1
1980	9.2	15.8
1981	6.0	24.0
1982	7.0	21.0
1983	5.8	20.2
1984	8.0	19.0
1985	7.6	26.4
1986	5.4	31.6
1987	7.0	35.0
1988	8.5	35.5
1989	10.2	39.8
1990	7.9	33.1
1991	10.1	31.9
1992	13.3	36.7
1993	12.6	38.4
1994	11.8	39.2

Source: *Cedigaz.*

LNG

Libya became involved with LNG in the early days of the industry. In 1964, Esso decided to profit from the associated gas produced in its oilfields and secured contracts for the sale of LNG to Snam, a subsidiary of Italy's national oil company ENI, and with Catalana de Gas, Spain's largest lighting, power and gas utility. A year later, it began the construction of what was then the world's largest two-train liquefaction plant, with a capacity of 3.5–3.9 bcm per annum at Brega on the Sirte Basin coast.[7]

The contract with Snam was for 2.4 bcm per annum for 20 years and included construction of a regasification terminus at La Spezia, near Genoa. The Catalana contract, for 1.1 bcm per annum, was to run for 15 years and included the construction of a regasification terminus at Barcelona. Catalana created a new company, Gas Natural, to oversee the Spanish gas industry; Enagas took over its LNG contracts. The cost to Esso was estimated at $320 million, but in the end this estimate was exceeded by $30 million. Approximately $196 million was set aside for the plant and the balance was earmarked for the construction of pipelines and other infrastructure and for four LNG tankers.[8] The tankers were costed at $78.9 million.

Despite its great promise, the project was never a happy one. To begin with, construction costs escalated and technical difficulties and fires delayed completion of the Brega plant. When Esso was finally ready to proceed, the Libyan government of King Idris had fallen. The story goes that Esso was actually loading its first LNG shipment in the summer of 1970 into one of its four purpose-built tankers when the Qaddafi government refused to permit exportation on the grounds that the Esso f.o.b. price, on which government revenues were based, was too low. Esso claimed that the Idris government had agreed that it could base its taxable income on a transfer price from Esso Libya to an affiliate, Medstan, which was then free to sell to Snam and Enagas at a different price.[9] The government refused to recognize 'the artificially reduced price at which the gas is sold to Medstan as the true f.o.b. price on which the government's tax and royalty income should be based'.[10] The original Medstan price was in the region of 20.6 cents/mBtu. The government argued that this low price would give it no income on gas exports and demanded a considerably increased price.

In March 1971, the government signed an agreement with Esso for a f.o.b. price of 34 cents/mBtu for exports of LNG to Enagas in Spain. The agreement contained a clause allowing for gradual increases in price to offset the effects of future inflation in Europe. In July, it signed a second agreement on a f.o.b. price to Snam of 34.5 cents/mBtu, also linked to an inflation index, with the first automatic inflation escalation of 2.1 cents/mBtu set for January 1972. The government also took an option to buy up to 50 per cent participation in Esso's Brega LNG plant, an option which it did not pursue for seven years.[11]

By 1973, the Brega plant was working at an average of about 80 to 90 per cent of capacity and Esso had contracted to buy gas produced by Oasis and Amoseas in their associated fields to make up the necessary volumes for export. A year later, the picture had changed. In October 1974, the Libyan government invoked the inflation clauses in its agreements with Esso and issued a decree raising the f.o.b. LNG price to all customers to $1.62/mBtu, a price related to what Algeria was prepared to accept for new gas supply contracts.

Both Snam and Enagas refused to pay the higher prices passed on to them. Exports ceased and the Brega plant closed. Several weeks later, Enagas agreed to pay the higher f.o.b. price as an 'interim measure' and the plant started up again at a level which enabled it to meet export commitments to Spain. An agreement on a complex pricing formula with Snam was not reached until March 1975 with the f.o.b. price on half of the LNG supplied raised to $1.62/mBtu and the price rise for the remaining half not due to occur until 1977.

New disputes over LNG pricing arose in 1979, when crude oil prices increased dramatically and OPEC members openly discussed the idea of linking the price of natural gas with that of crude oil. In January 1980, Libya called for a f.o.b. price increase to $3.45/mBtu, to which Enagas agreed. In August, the Libyan oil minister announced that events had overtaken that settlement and a new deal was needed. He said that Libya was co-ordinating its pricing stand with Algeria, which was seeking over $6/mBtu, the f.o.b. parity with crude. Both Enagas and Snam refused to accept this price and, in the end, neither Algeria nor Libya succeeded in imposing prices on their customers anywhere near this goal.

In the middle of this dispute over prices, and perhaps partly because of it, Esso announced in November 1981 that it was withdrawing from all Libyan oil and gas operations, a decision, the company said, based on commercial considerations. Sirte Oil then took over the running of the Brega LNG project and the selling of its production.

As Table 9.5 shows, Libyan LNG exports to Italy fell sharply in 1980 and then virtually ceased except for an occasional spot trade deal. Snam refused to accept high Libyan prices because it had domestic production and imports from Russia and the

Table 9.5: LNG Exports. Billion Cubic Metres Per Annum

Year	*Italy*	*Spain*
1971	0.6	0.6
1975	2.24	0.89
1976	2.59	0.94
1977	2.61	1.01
1978	2.48	0.97
1979	2.09	0.92
1980	1.35	0.59
1981	0.01	0.74
1982	0.02	0.80
1983	0.03	0.74
1984	0.34	0.77
1985	0.28	0.76
1986	-	0.86
1987	-	0.80
1988	0.19	0.87
1989	0.28	1.15
1990	-	1.24
1991	-	1.58
1992	-	1.84
1993	-	1.60
1994	0.05	1.43

Source: *Cedigaz*.

North Sea and, after the Trans-Med pipeline was completed in 1983, ample supply of Algerian gas. Libyan LNG exports to Spain were generally lower between 1980 and 1988 but continued because Spain had no alternative for gas supply at that time.

In Table 9.6 it can be seen that prices to Enagas in Spain gradually decreased between 1982 and 1988, when they reached $1.75/mBtu f.o.b. By this time, decreasing f.o.b. crude oil prices had dampened the enthusiasm of OPEC gas producers for matching LNG prices to those of crude, with *ad hoc* and interim price agreements becoming more common. After 1988, Libya quoted LNG prices on a c.i.f. basis, which came out approximately 50 cents higher than the f.o.b. basis. In the early 1990s, the price was generally around $2.50/mBtu except during the first half of 1991, when it rose briefly to $4.95/mBtu. The Libyan government eventually made an arrangement with Enagas which

Table 9.6: LNG Prices to Enagas, Spain. Dollars Per Million Btu

1982	4.50 f.o.b.
1983	3.95 f.o.b.
1984 - May	3.45 f.o.b.
1985 - March	3.35 f.o.b.
1985 - September	3.15 f.o.b.
1986 - April	2.66 f.o.b.
1987	n.a.
1988 - October	1.75 f.o.b.
1989 - October	2.62 c.i.f.
1990 - January	2.47 c.i.f.
1990 - April	2.45 c.i.f.
1990 - July	2.15 c.i.f.
1990 - October	2.58 c.i.f.
1991 - January	4.35 c.i.f.
1991 - April	3.95 c.i.f.
1991 - July	3.15 c.i.f.
1991 - October	2.75 c.i.f.
1992 - January	2.55 c.i.f.
1992 - April	2.40 c.i.f.
1992 - July	2.50 c.i.f.
1992 - October	2.70 c.i.f.
1993 - January	2.80 c.i.f.

Source: *Cedigaz*.

allowed LNG prices to vary in conjunction with the price of a basket of crudes.

Export Expansion Plans

For a while in the 1980s, the Libyan government was not overly concerned with faltering LNG exports as its ambitious development plans called for the increased use of natural gas in Libya. By the latter part of this decade, however, it began to hanker after increased gas exports and the hard currency these would bring. It was apparent that LNG sales to Italy, given the existence of the Trans-Med pipeline, were unlikely to start up again, at least in any quantity.[12] Spain, however, continued to be a reliable customer and on 5 June, 1990, the government signed a new contract with Spain intended to build up sales to 2 bcm per annum by 1994 and to run until 2008.[13]

The government felt it was necessary to seek new customers and it started discussions with Turkey, Greece, Yugoslavia, Belgium and Portugal regarding LNG purchases. A provisional agreement in 1988 with Turkey's state agency Botas for the sale of 1.5 bcm per annum over 25 years did not materialize. Shipments were delayed pending completion of a terminal at Marmara Ereglisii, and in the meantime, Botas was reported to be receiving Algerian LNG and to have made arrangements with other LNG suppliers, including Australia. In 1991, the government had announced that it was 'preparing a new model contract to attract foreign investment in the gas industry'.[14] In this it may have been influenced by a fear of increased Algerian gas exports following the Algerian law in that year giving foreign companies an entry into the Algerian gas industry formerly exclusively reserved for Sonatrach.

A major problem which the government needed to solve in order to increase LNG sales, however, was refurbishment of the Brega plant which was built to liquefy a high level of LPG along with the LNG; most Libyan gas is especially rich in LPG. Snam and Enagas solved the LPG content problem by building, at Barcelona and La Spezia, fractionation plants in front of their regasification plants which could handle LNG with a LPG content sometimes as high as 25 per cent. None of the potential customers with which the Libyan government was negotiating in the early 1990s had regasification plants with LPG separation facilities on the scale needed for high-calorie Libyan LNG, nor did they express any desire to construct these.

In 1991, the French company Technip was appointed to design the modification of the Brega LNG plant and bids were expected to be requested from engineering companies for this work in late 1993. The refurbishment of the Brega plant, however, was delayed by financial problems and by the restraints posed by UN and US sanctions. It had not been completed by the end of 1995. Another reason for the delay is reported to be the preference of some members of the Libyan government and the NOC for the proposed Agip pipeline intended to take natural gas from the NC-41 offshore field to Italy. They see this pipeline as a preferred option to increased LNG exports. They would like to have a central gas processing and export terminal near Tripoli to handle not only NC-41 gas but also gas to come from the Al Wafaa field to the south and gas sent along the pipeline from Brega. Advocates of this project – whose realization depends far more on Agip plans for all of its Mediterranean operations and Italian government policy than on Libyan preferences – feel that the investment needed to upgrade the Brega LNG plant would be better spent on preparing for pipeline exports.

At the end of February 1996 it was reported that the government was close to finalizing an agreement with Agip for the processing and export of natural gas from the NC-41 offshore field and the onshore Al Wafaa field. The dry gas would be exported to Italy via a pipeline.

LPG

Libya also produces some LPG for commercial purposes. Both Occidental and the BP/Hunt partnership built gas liquid extraction plants at their oilfields and Occidental constructed a pipeline to carry LPG from its Intisar field to Zueitina where exports began in 1972. The BP/Hunt partnership, however, returned the liquids they extracted to the main crude flow to raise its gravity and reinjected the lean gas into its Sarir fields.

Libyan LPG exports, which go primarily to Europe, are small and erratic, as Table 9.7 indicates. Its principal markets are

Italy and the Netherlands, although France took several shipments in the mid-1980s and occasional sales have also been made to Belgium, Germany, Greece, Portugal and Spain.

Table 9.7: LPG Exports to Europe. Metric Tons

1986	91,000
1987	35,000
1988	6,000
1989	16,000
1990	39,000
1991	10,000
1992	30,000
1993	42,000

Source: *IEA Gas Information*, 1987–89, 1989–91, 1993.

Future Prospects

It seems likely that Libya possesses ample undeveloped and undiscovered gas reserves. Elf Aquitaine has reported gas finds in its offshore block NC-137 which is close to the rich NC-41 gas field and the joint Libyan–Tunisian 7 November Block. All of this offshore gas is in the vicinity of the Tunisian Miskar gas field under development and these discoveries had not been appraised by 1995. Sirte Oil has reported recent finds at Meghil, 50 miles south of Attahadi, and Veba has struck gas in block NC-72, at least 100 miles south-west of Meghil.

Most of the oil discoveries to date have been found in anticline reservoirs and many geologists are convinced that there is more oil (and gas) to be found in stratigraphic reservoirs. Finding such reservoirs, however, is difficult and expensive. A considerable financial investment will be required to find and develop Libyan gas reserves and an even greater investment will be needed to build a larger export system. The indications in the early 1990s were that the Libyan government realizes that it cannot establish a significant export capacity for its natural gas without foreign capital and expertise.

Notes

1. The report did not include the production and export of associated gas from the Bouri field in the final project, as was originally planned, perhaps because this gas is extremely sour.
2. The name of this block commemorates the date of Zein al-Abidine Ben Alithe's assumption of the Tunisian presidency.
3. *Middle East Economic Survey*, 33/15, 15 January, 1990, p. D4, citing Tayeb Ounada and Mousa Ismail, 'Status of the Gas Industry in the Arab Countries and the Prospects for its Development', OAPEC Technical Affairs Dept.
4. It was not clear whether or not this was in operation in 1995.
5. *Cedigaz News Report*, 28 January, 1994, p. 11.
6. The recoverable associated gas reserves from Sarir are estimated at about 600 bcf. *Cedigaz News Report*, 4 May, 1990.
7. Bechtel was the engineering contractor for the Brega plant and Snam was the construction contractor. The process installed was the Air Products and Chemicals Inc. Multi-Component Refrigerant system, a combination of a pure refrigerant cascade process and a mixed refrigerant cycle.
8. Esso also planned to build a plant at Brega to extract gas liquids – LPG and naphtha – from natural gas and to export these, but it is unclear whether this plant was ever constructed.
9. *Petroleum Intelligence Weekly*, 8 June, 1970, pp. 5–6.
10. *Middle East Economic Survey*, 13/37, 10 July, 1970, p. 2.
11. In 1980, Esso agreed to Libyan majority participation because a lot of investment was needed for a full re-vamping of the plant. *Petroleum Intelligence Weekly*, 26 May, 1980, p. 11.
12. In late 1994, there were reports of discussions with Snam over a new contract for the sale of 1 bcm per annum to Italy, but these had not come to anything by the end of 1995.
13. The 1994 deadline was not reached, and, as Table 9.5 shows, LNG sales to Spain fell in 1994 to 1.43 bcm from 1.60 bcm in 1993.
14. *Petroleum Economist*, November 1994, p. 15.

10 PETRODOLLAR INVESTMENTS

A World Bank mission which visited Libya in the late 1950s noted that the discovery of oil rarely provides an easy or complete solution to the problems of economic development. This turned out to be an accurate forecast. Like other countries with virtually no natural resources other than oil, Libya found it very difficult to absorb rapidly increasing government revenues in ways that benefited its overall economy. It became a clear example of what has been termed a 'rentier state', that is to say, a country that receives revenue from the production and export of a raw material that has little or no connection to economic growth.

As Table 10.1 shows, Libya experienced a sudden enormous rise in GNP per capita from an 1951 estimate of $35 to more than $5600 in 1974 and close to $12,000 in 1980. Although the government drew up a series of Five-Year Plans and poured ever-increasing funds into the agricultural and industrial sectors,

Table 10.1: Libyan GNP at Current Market Prices. US Dollars Per Capita

1951	35	1977	7291
1960	258	1978	6886
1961	276	1979	9107
1962	369	1980	11706
1963	514	1981	9922
1964	745	1982	8973
1965	967	1983	8304
1966	1194	1984	7067
1967	1357	1985	6420
1968	1849	1986	5563
1969	2028	1987	6132
1970	2060	1988	5122
1971	2272	1989	5647
1972	2542	1990	6013
1973	3342	1991	6124
1974	5629	1992	5421
1975	5254	1993	4714
1976	6475	1994	4256

Source: *OPEC Annual Statistical Bulletins.*

the country's fundamental economic problems – a deficiency of arable land, water, minerals and human resources – remained unchanged. Stagnation in agricultural development, partly as a result of the flow of the rural population to the cities to seek the benefits of wealth and modernization, led to increased reliance on imported food. Industrial performance, except in the oil sector, improved only slightly. Where there were gains was in the social sector, in health, education and, to a lesser extent, public services such as electrification and transport.

The problem which the Libyan government faced was how to distribute its oil revenues to stimulate long-term economic activity outside of the petroleum sector. It could subsidize imports, but the effect would be to discourage rather than foster economic independence. It could (and did, to some extent) invest abroad, but the consequence of this, when successful, would be mainly the creation of more government income. It could support secondary industries which manufacture consumer goods, but the internal market for these was small in a country with a population of only four million and the prospects for such goods being able to compete in markets abroad with mass-produced goods are small.

What it did was to invest its oil revenues in five main areas over the years, with varied emphasis:

i. The oil industry and its offshoot, petrochemicals.
ii. Agriculture, especially water resources.
iii. Other industry, with emphasis on iron and steel.
iv. Physical infrastructure, such as dams, power plants and roads.
v. Social infrastructure, such as education and health.

Throughout the 1970s the government found that projects in which it was investing, with a few exceptions, were unable to absorb all the funds available. After 1980, the reverse became true. Decreasing oil revenues left the government short of funds to cover the needs in all its investment areas, especially after it embarked on the Great Man-Made River project, with the result, as noted previously, that the oil industry, in particular, was starved of needed investment.

Dependence on Foreign Aid

As noted in Chapter 1, when it became independent in 1951, Libya was one of the poorest nations in the world, with an estimated per capita income of $35 for its population of slightly less than one million. There were no known mineral deposits, except for large but remote iron ore deposits in the Fezzan, although there was hope for petroleum reserves. Manufacturing was limited to tanning and the production of various types of leather goods. Salt was produced and there was some food-processing and a few small workshops. There was very little installed electric power capacity. Some 95 per cent of the country was desert and only about half of the remaining 5 per cent was suitable for cultivation. There were no perennial flowing streams or rivers and access to water in underground reservoirs was often difficult as a result of tribal land tenure which also complicated farm tenancy. The large-scale state agricultural programmes which Italy had sponsored during its interwar occupation of Libya ended with the outbreak of the Second World War without having achieved any notable success.[1] The countryside suffered much war damage in the north-east province of Cyrenaica.

Adrian Pelt, the United Nations Commissioner given the task of preparing Libya for independence, reported in 1950 that the country had 'a marginal agricultural economy basically handicapped by inadequate rainfall and poor soil' and that the 'crop surplus for export was small and irregular, of low quality and high price'. He went on to note:

> A lower per capital income than any other country in the Middle East, a very high birth rate ... marginal or sub-marginal land, low and 'capricious' rainfall and frequent droughts, absence of minerals or fuel and, above all, of skill and education.
>
> The deficit in the public finances of nearly £2 million cannot be met by increasing the taxes on an estimated $35 per capita and subvention is therefore necessary to preserve the present standard of living. The adverse balance of trade ... cannot be righted by a reduction of imports without yet further reducing the level of the

> economy; and this, too, must therefore be made good by external aid.[2]

Once Libya was declared independent, Britain, the United States and to a lesser extent France, were quick to offer foreign aid along with others, including Egypt. The motives of most did not stem from charity. The increasing threat of the Cold War made secure bases in friendly nations in the Mediterranean and the Middle East a defence priority for the West. Faced with the impending British withdrawal from the Middle East, the Algerian war of independence and the Moroccan call for complete evacuation of all foreign troops, Libya seemed a good investment. The Western allies gave assistance in the form of budgetary grants-in-aid, contributions to development schemes and spending on military projects. In 1952, the foreign assistance of Britain, the USA and France amounted to approximately £L (Libyan pounds) 8.6 million, compared with £L5.5 million in 1951 and £L3.8 million in 1950. In addition, Libya received another $1.5 million from the UN Technical Assistance Programme. Per capita, this sum was larger than UNTAP's investment in any of the 90 countries and territories where it was active.

Intense foreign development planning was foisted on Libya from 1948 until 1951, while it was a ward of the United Nations, and continued while it remained financially dependent on the West until the mid-1960s. The World Bank drew up a development plan which focused on education, agriculture and public works, but neglected industry. The several foreign agencies which set up offices in Tripoli had their own agendas as to how foreign aid should be spent.[3] Some of their programmes were unfortunate choices. The extensive planting of citrus, tomato and peanut crops which needed much water was disastrous, as it drew down the water table along the coast to such an extent that seawater intruded. Despite heavy investment in the agricultural sector, food imports more than doubled between 1954 and 1962.

Development Plans under King Idris

Once oil was discovered, the Idris government made plain its intention to use oil revenues to improve the Libyan economy. It was aware that economic development of the country would depend on the progress of agricultural expansion, on the construction of an adequate physical and social infrastructure and on fostering new activities in the manufacturing and services sectors. It issued a decree that 70 per cent of all sums received on account of oil royalties and taxes must be allocated to development. (The remaining 30 per cent were to be divided equally between the province containing the producing oil concession and the federal government).

The 1963 constitutional changes which abolished the country's original federal structure and established a unitary state led to central concentration of development planning. There was a National Planning Council chaired by the prime minister to formulate planning policy and a Ministry of Planning and Development to administer development plans. The first Five-Year Development Plan for 1963–68, with declared aims of raising the standard of living of the population and laying the foundation of future economic growth, was basically an enumeration of projects suggested by foreign development planners. The largest proportion of the funds (22.9 per cent) was to go for public works projects. Agricultural programmes focused on land reclamation and irrigation schemes. These were to receive 17.3 per cent, transport and communications 16.2 per cent, education 13.2 per cent, and industrial development only 4.1 per cent.

This programme was overtaken by economic events with oil revenues far exceeding earlier expectations. By the mid-point in the plan period, there were more funds available for investment in one year than had been estimated for the entire five years. By the end of the five years, expenditure on development had totalled £L298 million, considerably more than the £L170 foreseen in 1963, but only about 20 per cent of the amount received as oil revenues in these years – thus falling far short of the 70 per cent stipulated by law. Some of this shortfall was due to the reduction and final end of foreign assistance, as oil revenues were needed to meet ordinary government expenditures previously covered by these funds. In addition, large sums were spent on defence and on aid to other Arab states as a result of hostilities with Israel. Actual spending on public works absorbed more funds than budgeted, while agriculture and education received less. Some 10 per cent was spent on housing, a sector which had not been included in the original plan budget.

The results of all this investment were disappointing, with low annual rates of growth of the agricultural and industrial sectors. In addition, increased oil revenues caused other problems for the Libyan economy. Inflationary pressures developed, with the cost of living index rising and food prices increasing by 70 per cent between 1955 and 1967. There was also a drastic fall in agricultural production to less than half of its level in 1958, partly as a result of an exodus of agricultural workers to the towns, with agricultural labour as a proportion of the total work force falling from approximately half in 1963 to less than one-third by 1969. A shortage of manpower was added to the problems of poor soil, erratic and insufficient rainfall and an unpredictable climate. Libya, which had been a net exporter of agricultural produce, became a net importer, with domestic agricultural production only meeting 40 per cent of home demand.

The government was aware of these problems and announced its intention to counteract inflation, stop the drift from the agricultural sector and increase domestic agricultural production. It attempted to support agricultural prices at levels deemed to be conducive to increased production and issued long-term agricultural loans free of interest. Its three major development projects, however, ran counter to these goals. These were to construct 100,000 housing units in Libya's three cities, to complete the construction of a third city and new capital, Beida, and to construct a coastal highway linking Tripoli to Tunis. It drafted a second Five-Year Plan for the period 1968–73 but could not agree on its substance; meanwhile, the structure of the first plan was extended into a sixth year. The proposed second plan was similar to the first and focused on investment in infrastructure, with almost half of the funds earmarked for public works, housing and transport and another 17 per cent on education, health and other social needs. Agricultural investment was reduced to 13.1 per cent, and industrial investment raised to 7.9 per cent.

Economic Development Plans of the Revolutionary Command Council

Although the Revolutionary Command Council, when it took over in September 1969, scrapped the previous government's second Five-Year Plan, it followed the broad development goals set by the former regime, especially with regard to increasing agricultural output in order to achieve self-sufficiency in the production of basic foodstuffs. Its first development budgets were *ad hoc*, single fiscal year allocations; these were replaced by a Three-Year Plan for the period 1973–75 which focused on agricultural development. Like previous plans, this was an enumeration of projects which did not appear to be integrated into an overall, long-term development policy although the government established a Technical Planning Authority to manage the investment of the nearly $6 million allocated.

The new planners were filled with optimism at first. In 1970, Major Abd al-Salem Jallud, a prominent member of the Revolutionary Command Council, pronounced that Libya

> would be transformed within a period of two years into a country whose main wealth derives from agriculture and industry ... Within one year Libya will become almost self-sufficient for cement and in less than two years we shall have factories for glass, dairy products, medicines, etc. ... We should consider oil revenues as foreign loans to promote the country's agricultural and industrial development plans.[4]

The government soon faced the reality of constraints on agricultural and industrial development. Within three years, Major Jallud had conceded that 'the basis of a modern economy in Libya will be the oil industry and its offshoots' and said that the government intended to create a diversified oil industry including downstream and petrochemical projects.[5] There would still be a considerable effort to diversify the sources of national income by exploiting resources other than oil, he added, underlining the importance of the discovery of further iron deposits.

This statement signalled increased emphasis on industrial projects which the government proceeded to classify into major, intermediate and minor categories. It declared its intention of keeping control of major projects, which it defined as those dealing with the processing of important natural resources. It also intended to be involved in intermediate category projects, described as those involving the processing of important goods for local consumption, including agricultural products, building materials and goods produced from imported raw materials. Foreign investment would be permitted in major and intermediate projects if such investment would introduce advances in technology and would result in exports. Public sector financial participation in projects in both of these two categories, however, would be at least 51 per cent. Projects classified as minor, defined as those concerned with the manufacture of consumer goods of lesser importance for the domestic market, would be left entirely to private sector investment and control.

One of the problems which the government was unable to solve with regard to industrial projects in particular was the limitation of a small population. The demand for additional labour and expertise in this sector had to be met largely by foreigners. In 1975, immigrants were providing 42 per cent of the country's unskilled labour, 27 per cent of skilled and semi-skilled workers, 35 per cent of its technicians and 58 per cent of management and supervisory staff. By 1980, some 13.6 per cent of the total population was non-Libyan and more than 34 per cent of the labour force was non-Libyan.[6] The foreign population included not only Egyptians and Tunisians but also workers from western and eastern Europe.

The 1981–85 Five-Year Plan considerably increased the

Table 10.2: Libyan Development Budgets. Major Sector Allocations. Percentages

Plan	*Industry*	*Agriculture*	*Housing/ Utilities*	*Communication/ Transport*	*Education*	*Oil*
1970–71	10	25	15	13.6	5.7	-
1971–72	10	16.8	15	13	10	7
1973–75	12.4	20.9	14	9.4	8.9	8.5
1976–80	15.4	18.1	18.9	9.1	6.7	9.3
1981–85	23.1	18.2	21.8	12.3	5.9	1.2

Sources: *Middle East Economic Survey*, various issues; COMET Report.

investment share for industry. Despite improvement in average living standards, Libya's fundamental economic problems remained when this plan ended. By this time, oil revenues and GNP per capita had fallen and development strategy changed. The government did not publish another development plan for over a decade. During this period, it directed investment primarily into three large projects, petrochemicals, the Great Man-Made River and the iron and steel works at Misurata.

Oil Sector Investments

The early development plans did not include specific allocations for the oil and gas sector as these were chiefly under the control of foreign oil companies. When the Libyan National Oil Company became active, funds were set aside for its work in this sector. The investment proportion in the budget for oil and gas gradually increased from 1970 to 1980 and then it was dramatically cut back in the 1981–85 Five-Year Plan. Some of this reduction, however, reflected the reallocation of funds for the petrochemical industry from the oil budget to the industry budget.

The breakdown of specific allocations within the oil and gas sector in 1971 and 1978 (see Table 10.3) shows an impressive increased emphasis on the development of natural gas resources and a marked decrease in expenditure on refining.

Although figures are lacking for oil and gas sector investments after 1985, it is generally believed that the funds which the government has made available to the industry in this period have been insufficient for expansion and probably even for maintenance.

Petrochemicals

Plans for a Libyan petrochemical industry surfaced early as the Idris government saw this as a way of using the natural gas which was then being flared during the extraction of oil from associated fields. (Ironically, when it was finally built, the principal Libyan petrochemical industry in Ras Lanuf depended on refinery-produced naphtha as its feedstock, and natural gas was used only in the fertilizer plants in Brega.)

Table 10.3: Allocation of Oil and Gas Sector Expenditures. Per Cent of Total Investment

Area	*1971*	*1978*
Refining	34	18
Gas Projects	2	50
Exploration and Production	19	17
Training	5	5
Western Libya Pipeline	-	8
Other*	40	2

* In 1971, this category included expenditure for the petrochemical industry, tankers, the establishment of a drilling and geophysical company, product storage tanks and non-specified; in the 1978 budget, this category was not broken down.

Source: *Middle East Economic Survey*, 15/41, 4 July, 1972, p. 9; 21/25, 10 April, 1978, p. 7.

As in the case of refineries, the government turned to a foreign oil company to make the first move in building a petrochemical industry. In 1966, it persuaded Occidental to construct a plant deriving ammonia from natural gas in return for the award of a concession. The agreement was to be based on a 50/50 cost-sharing basis and the ammonia was intended partly for the manufacture of fertilizers for the domestic market and partly for export. Construction of the plant at Occidental's export terminal at Zueitina was scheduled to begin in 1968 but Occidental was unwilling to continue with the project when its feasibility study indicated a world surplus for ammonia and therefore a weak export market. The company argued that it was not obliged to build a plant which would not be economically viable. The government disagreed and commissioned its own feasibility study.

Brega Complex. No further action regarding a petrochemical and fertilizer industry was taken until 1972 when the Qaddafi government resolved the ammonia plant dispute with Occidental by agreeing that the company could construct a methanol plant instead, on similar terms. A National Methanol Company was set up as a 50/50 joint venture between Occidental and the NOC and a construction contract with a three-year completion date was signed with the West German company Friedrich Uhde. The plant was designed to have a production capacity of 1000 tons a day (t/d) of methanol and was to be part of the complex to be built in Mreisse, 18 miles south of Benghazi, and connected by pipeline to the Zueitina associated oilfield. Before construction on the plant began, however, the government decided to move the site from Mreisse to Brega, near the Esso facilities. In addition to the methanol plant, the plans for Brega included an ammonia plant, a carbon black plant and an urea plant, all to be up and running by 1976.

In the meantime, NOC had decided to go ahead with the construction of a 1000 t/d ammonia plant at the Brega site, following a feasibility study done by a French company which was more optimistic regarding the world market than the study commissioned by Occidental. It intended that 70 per cent of the ammonia produced would be converted into fertilizer at the future urea plant for domestic use and the rest exported in liquid form. The plant was to be the first step towards realizing the government's programme to make Libya self-sufficient in ammonia-based fertilizers. Built by Friedrich Uhde, it was in production by the end of 1977, operated and maintained by a British company, Coppas International.

The Brega methanol plant also came into operation at this time. The Mediterranean littoral was seen as the main market for methanol exports, but the National Methanol Company also hoped to sell on a spot basis in northern Europe despite the overcapacity of methanol production predicted by most experts for the world market by the early 1980s. In 1979, Occidental sold its interests in the National Methanol Company to the Libyan government for $9.9 million. In this same year, Foster Wheeler Italiana completed the construction of a 1000 t/d urea plant in Brega. The government commissioned a British firm to study the enlargement of Brega harbour and signed a joint-venture agreement with Ashland to construct a carbon black plant. It also signed a contract with the Italian ENI subsidiary, Snam Progetti, for a second 1000 t/d ammonia plant, with completion scheduled for 1981. When the first ammonia plant was badly damaged by an explosion in April 1988, Friedrich Uhde assisted in its reconstruction, which was complete by the beginning of 1991.

Ras Lanuf Complex. With the Brega site works well underway, the government focused on plans for the construction of a petrochemical complex at Tobruk in conjunction with the 220,000 b/d export refinery scheduled for this site. This complex was to include a 330,000 tons/year (t/y) ethylene plant using naphtha from the refinery as feedstock, a 171,000 t/y propylene plant and a 135,000 t/y butane plant. Before work began, however, the site of the export refinery and petrochemical complex was moved from Tobruk to Ras Lanuf at the suggestion of a Belgian consultant, Tractebel, which did a feasibility study for the government.

In 1982, the government established the Ras Lanuf Oil & Gas Processing Company (Rasco) to manage the refinery and petrochemical complex at Ras Lanuf. Rasco shortly thereafter signed a contract with Stone & Webster to design the 330,000 t/y ethylene plant which began production at the end of April 1987, operated by the Yugoslav firm of Hemijska Industria Pancevo (HIP).[7] By 1989, Ras Lanuf petrochemical plants were producing ethylene, propylene, butane and pyrolysis gasoline. It had already been decided to go ahead with the second phase of the petrochemical complex which involved units using ethylene and propylene as feedstocks. In March 1988, ten European and South Korean companies submitted bids for the design and construction of a polyethylene plant, a polypropylene plant and units to produce methyl t-butyl ether (MTBE), benzene and liquefied petroleum gas (LPG). In mid-1989, Rasco awarded a contract to Yugoslavia's Energoinvest and two other Yugoslav concerns, HIP and INA-Projekt, for the construction of units to produce butadiene, MTBE, butene-1 and benzene. Completion of this work was delayed by licensing difficulties due to US economic sanctions against Libya and by the Yugoslav civil war. As of late 1995, these units were not operating and construction of the low-density and high-density polyethylene plant and the polypropylene plant had been indefinitely postponed.

On a smaller scale, Libya built several chemical and petrochemical plants at Abu Khammash, near the Tunisian border in the early 1980s. A West German consortium, Salzgitter Industriebau, constructed these plants which have an annual capacity of 60,000 tons of PVC, 50,000 tons of caustic soda and 8000 tons of hydrochloric acid.

The Great Man-Made River Project

Shortage of water has restricted agricultural production in Libya to two narrow, fertile coast strips, the Jefara plain around Tripoli and the Jebel al-Akhdar plain in northeastern Cyrenaica. Despite the large quantities of development funds lavished on the agricultural sector from the early 1960s onward, agricultural productivity declined and the sector proved unable to absorb investment effectively. In the early 1990s, agriculture contributed only 5 per cent of GDP and food imports were averaging 20 per cent of total imports.

The early efforts of the Idris government to reach self-sufficiency in food production involved several expensive experiments that were economically and environmentally unsound, resulting in erosion of water resources and degradation of pasture land. Limited water resources continue to thwart attempts to expand the area under cultivation or to increase agricultural yield. The first major agricultural project involved the Kufrah oasis, some 600 miles south of the coast, where large underground water resources had been found. In its 1966 concession agreement, Occidental agreed to invest 2.5 per cent of its profits in the development of agriculture at Kufrah. These funds led to the establishment of an experimental farm project producing wheat and other irrigated grain crops which was taken over in 1970 by the Qaddafi government. US consultants suggested that the area should be devoted to intensive livestock raising of 250,000 Barbary sheep reared on alfalfa. Within a few years, groundwater levels had been seriously reduced and massive transportation and feeding costs made Kufrah-grown mutton far more expensive than imported mutton. In 1977, the Barbary sheep-rearing plan was cancelled and the project reverted to grain production on a reduced scale.

Another extensive agricultural development project of the Idris government with equally meagre results was a plan to grow wheat on the Jefara Plain and at Sarir in the south-east. Scant rainfall in these areas eventually led to the abandonment of most wheat growing and the return to cattle grazing.[8] More successful agricultural development projects of the Idris, and later the Qaddafi, governments, included the construction of irrigation dams and desalination plants, experimentation with different crops and the use of price controls to discourage the cultivation of water-intensive crops like peanuts, tomatoes and watermelons.[9]

The main thrust of the Qaddafi government, however, became the exploitation of underground water reservoirs. Libya contains two of the world's largest desert groundwater basins, one in the south-west Murzuq Basin and the other in the south-east Kufrah-Sarir Basin. Although a Swiss company, Electrowatt, first proposed bringing water from Sarir to the coast in 1974, the government only decided to go ahead with this idea in 1983 when it launched the Great Man-Made River Project with a cost estimate of some $25 billion. There were five stages, or phases, to the project. The first, which involved bringing water from the Kufrah-Sarir Basin reservoir to the north-east coast area was completed in 1991, supplying water for municipal use in Benghazi as well as to other towns and irrigation projects in Cyrenaica. The second phase, which involves bringing water from the Murzuq Basin reservoir to Tripoli and the north-west coastal area was on target in 1995 for completion in September 1996. Phase three will involve increasing the capacity of the north-east water supply, and the final two phases will involve extending coastal conveyancing lines to additional agricultural, domestic and industrial consumers. These could be delayed as there are differing views regarding the best use of the water brought from these desert aquifers. Some feel that it should be employed to extend the amount of land under cultivation, others feel it would be used better to increase productivity of land already under cultivation where acceptable rural infrastructure already exists.[10]

Iron and Steel

The Qaddafi government embarked on a major industrial project in the mid-1970s when it announced plans to construct two state-owned iron and steel mills, with capacities of 670,000 t/y and 650,000 t/y at Misurata, on the coast east of Tripoli. It was intended that in their first years, these plants would rely on imported iron ore; later they were expected to process iron ore mined in the Wadi Shatti deposits in the Fezzan, 435 miles to the south and brought to Misurata by a new railway network.

The cost of the project was estimated at $6 billion, a figure which included the construction of port facilities at Misurata as well as the construction of a power plant and a desalination plant.

The port facilities were completed by a consortium of Yugoslav companies by the end of 1978 and work subsequently began on the steel mills which were to be direct reduction electric arc plants using the Midrex system. October 1988 saw an initial output from these plants; by 1990 production was recorded as 492,000 t/y. Output increased to 650,000 t/y in 1992 and to 874,000 t/y in 1994. Iron and steel products were exported to Egypt in barter agreements which included an assortment of manufactured Egyptian goods such as aluminium ingots, vehicles and household appliances. LISCO, the state iron and steel company, awarded a contract in the early 1990s to a European consortium composed of two German companies led by Austria's Voest Alpine to construct a third direct reduction plant at Misurata with a capacity of 650,000 t/y. The justification for this project lies in the iron deposits in the Fezzan; however the iron for the plants was still being imported entirely in 1996, due to the lack of transport facilities from the Fezzan deposits to the coast.

Despite its heavy financial commitment to the Misurata complex and the Great Man-Made River, the Qaddafi government did not entirely neglect small industrial development schemes. It established LIGIC, the Libyan General Industrial Corporation, to coordinate planning for all state-operated industrial projects and to administer industrial agreements signed with foreign companies. LIGIC was instructed to draw up industrial development plans within the framework of the overall development strategy, obtain approval for projects included in these plans and take steps for their implementation either directly or in cooperation with others. It particularly favoured construction projects and approved so many contracts for these that Libya was reported to have the highest per capita consumption of cement in the world in 1974. It also approved investment in small-scale industries engaged in the manufacture of electrical machinery, batteries, footwear, textiles and clothing. In addition, it assisted industries involved in food-processing activities such as dairies, flour mills, refrigeration plants, canneries and fodder production companies. These efforts were very similar to those of the government under King Idris, although on a larger scale. The building of electricity-generating plants and the expansion of existing and new ports were also continued activities.

Recent Development Plans

The collapse of oil prices in 1986 put an end to formal development plans for almost a decade. In 1988, the government announced an austerity programme as a result of low oil prices and the effects of US boycotts. It curtailed investment in education and housing and concentrated even more on its major projects. Infrastructure investment dropped precipitously, even to the extent of neglecting maintenance, especially in the transport sector. Controls over imports were increased by strictly enforced requirements for import licences for most non-food items and limits on the latter food imports. In 1993, the government unexpectedly announced a three-year development plan for 1994–96. It did not issue a breakdown of investment figures but set out its aims as:

i. Completing agreed projects such as the Great Man-Made River and a proposed railway link to Egypt.
ii. Improving productivity and profitability in agriculture and manufacturing.
iii. Increasing infrastructure in remote areas and expanding the housing sector.
iv. Enlarging commodity storage capacity.
v. Increasing reliance on Libyan companies and Libyan labour.[11]

It is doubtful if many of these goals will be met within the time period stipulated.

Investments Abroad

In the 1970s, the government decided to invest abroad some of the surplus funds which the internal economy could not absorb and proceeded to build up large holdings of foreign assets. The net foreign assets of its central bank and commercial banks were

$2.4 billion in 1973; by 1982 these had increased by $22.2 billion.[12] In 1972, the government established the Libyan Arab Foreign Bank (LAFB) as an overseas Libyan bank fully subscribed by the Central Bank of Libya for the purpose of undertaking financial and banking operations outside Libya. Although a significant portion of LAFB's available funds were placed in short-term investments, it was also involved in equity investments, loans, including a $500 million loan to ENI in Italy, bonds and purchases of real estate in both Italy and Malta. It participated in consortium banks and became a minority shareholder in smaller banks, typically those concerned with trade financing. It also set up several joint banks in the Third World, in Mali, Uganda, Togo, Niger, Chad, Mauritania and Burkina Faso.[13]

In 1977, Libya made a major industrial investment in Europe with its purchase of a 9.5 per cent, later increased to 15 per cent, stake in Fiat; the sale of this holding in 1986 realized a large profit. As part of the original deal, Fiat agreed to build a $50 million assembly plant for trucks and buses in Libya. This was to be a forerunner of similar industrial investments intended to give Libya access to technical know-how in a range of different industries which were to be channelled through the Libyan Arab Foreign Investment Company (LAFICO). This programme was never realized, however, and almost all of Libya's subsequent equity investment overseas was connected to its petroleum industry. In 1988, the government established Oilinvest to manage the overseas petroleum investments discussed in Chapter 8. The shareholders of Oilinvest were LAFB, LAFICO and NOC.

LAFB also maintained a substantial amount of international currency reserves in varied short-term and long-term arrangements in a veritable web of financial institutions.[14] In 1992, Libya transferred many of its international financial holdings out of Europe in advance of expected UN sanctions which threatened to freeze such assets. It is thought that most of these transfers went to banks in countries with commercial interests in Libya, as well as to traditional centres of discreet international banking such as Switzerland. Arab banking sources reported that $6.5 billion had been transferred. Sources which still reported Libyan holdings were said to have $6.2 billion at the end of 1992.

In addition to investments, Libya also expended funds abroad through politically motivated loans and aid. In its first few years, the Qaddafi government spent large sums in efforts to establish a unified Arab and Muslim world of which it saw itself as a leader. When loans and gifts to other Arab nations were received not with gratitude but with resentment, Libyan aid of this type diminished. After 1974, the government directed more of its aid towards sub-Saharan African governments.[15] The principal beneficiaries of this policy were Niger, Chad, Equatorial Guinea, Zaire and Uganda. Some aid also went to Mozambique and the Malagasy Republic. Financial considerations were always involved in the allocation of these funds and Libya preferred direct involvement in projects which seemed likely to guarantee a return on its investments rather than open-ended loans to governments. As a result, it sponsored a number of joint ventures in Africa, with a Libyan share of at least 51 per cent.[16] These joint ventures were concentrated on the agricultural, financial and mining sectors; a well-publicized one involved copper mining in Zaire.

The government also made loans and investment outside of Africa. These included participation in an Adriatic oil pipeline construction project with Yugoslavia, aid to South Yemen and loans to Malta, Pakistan, Malaysia and Turkey. Investment tended to focus on projects in western Europe after the oil price collapse in the mid-1980s.

Accomplishments

It is difficult to ascertain the results of the investment of oil revenues in other sectors of the economy as there are few reliable public sources of statistics on the Libyan economy available in recent years. The International Monetary Fund and the World Bank, for instance, carry no data on Libya after the mid-1980s, when diplomatic relations were severed. Some data are published by the UN, especially its Economic Commission for Africa, but the sources of these are not referenced. The British Economist Intelligence Unit also publishes unreferenced data as does Barclay's Bank and the Arab Banking Corporation.

As noted earlier, Table 10.1 shows that GNP per capita increased rapidly up to 1980 and then fell back, abruptly at first

Table 10.4: Libyan Social Statistics

	1970–75	*1990*
Average Annual Population Growth	4.2%	3.6%
Life Expectancy	52.9 Years	63.1 Years
	1982	*1991*
Urban Population	60%	71%
Population Engaged in Agriculture	4.2%	2.7%
Population Engaged in Industry	7.4%	7.9%
	1973	*1990*
Illiteracy Overall:	60%	n.a.
Women	n.a.	42.6%
Men	n.a.	14.5%
	1960	*1989*
Infant Mortality	160 Per 1000 Live Births	78 Per 1000 Live Births
	1965	*1990*
Doctors Per Capita	1 Per 3860 Persons	1 Per 500 Persons

Sources: UN Economic Commission for Africa, 'UN African Statistical Yearbook, North Africa 1990/91'; EIU Country Profile 1994–95 Libya; COMET Report; UNDP, *Human Development Report*, 1995.

and then more gradually, with a short-term increase in 1990–91 as a result of the invasion of Kuwait by Iraq. Other data are given in Table 10.4, but these require some interpretation.

The overall figure for population growth, for instance, has not been broken down into growth of the Libyan and of the non-Libyan population. This was estimated for the period 1980–85 as a growth of 3.9 per cent for Libyans and 4.5 per cent for non-Libyans.[17] The decrease given in the table for 1990, therefore, may reflect a drop in non-Libyan population rather than a slowdown in the average annual population growth of Libyans.

Once oil revenues began to flow into the government and development programmes began, Libyans flocked to Benghazi and Tripoli and other urban centres in search of better living conditions and employment. This trend has continued unabated in recent years, as Table 10.4 shows, with some 71 per cent of the population in urban areas in 1991. The effect of this migration is reflected in the decrease in the number of people employed in the agricultural sector. The percentage of the workforce engaged in agriculture in 1980, for instance, was 19 per cent; five years later it had fallen to less than 17 per cent. In 1992, agriculture contributed only 5 per cent of GDP, with food output continuing to decrease. Using an index of 100 for the years 1978–81, agricultural production peaked in 1982 at 128 and had fallen to 80 by 1991. These figures suggest that development in the agriculture sector has not been successful despite continued investment.

Unfortunately, there has not been sufficient industrial growth to absorb a significant proportion of the able-bodied who left the land. The oil and gas industry is not labour-intensive; throughout the 1980s, for instance, it employed less than 1 per cent of the population. The most important employer in the industrial sector is the construction industry, which accounted for 69 per cent of industrial labour in 1980 and 65 per cent in 1985. Many of those employed in construction, however, are non-Libyans.[18]

Allocations and actual expenditure for education and health have declined since 1970, relative to investments made earlier, presumably due to the completion of basic infrastructure programmes. Although the data in Table 10.4 show considerable improvement in these two sectors, they raise some doubts because different international sources such as the UN and the World Bank vary widely in their estimates of illiteracy rates in 1990. Despite government efforts to overcome this problem, one would expect continued illiteracy in rural areas and in the older generation. On the other hand, it is worth noting that the Libyan population is unusually young; it was estimated that 54 per cent of all Libyans in 1991 were under the age of 19. Comparing literacy rates with countries with a much higher average population age, therefore, is misleading.

There is no doubt that the government has invested a great deal in education from the elementary age level upwards, including education for women. The first Libyan university was established in Benghazi in December 1955; later one was established in Tripoli and later again, a technical university was opened in Beida. Education is apparently free through university levels and there is a notably high percentage of female students in the total university student body.

With regard to health, the decrease in infant mortality shown in Table 10.4 is impressive and is at an internationally respectable level. On the other hand, life expectancy is still low, at 56.6 years for men and 60 for women in 1990. The increase in doctors per capita is also notable, although many of these doctors, at least for many years, were foreigners. In 1980, for instance, only 10 per cent of the 4300 doctors in the country were Libyans.

One distinct achievement has been a large increase in installed electricity capacity and a consequent increase in electricity production from 6000 million kilowatt hours in 1982 to 19,000 million in 1990. In 1993, the government was reported as having signed a $1 billion contract for the installation of additional capacity. Construction of housing appears to have peaked in 1981 and the construction of roads, which was a priority in the first decades after independence, slowed considerably during the 1980s. There are very few, if any, functioning railway lines.

It would be wrong to surmise that the Libyan government frittered away its oil revenues and was therefore to blame for the fact that the country's economic problems have not been solved. Its efforts, for the most part, were well-intentioned but unsuccessful, partly because of a lack of technical and human resources but mostly because of the scarcity of resources which could be made productive. As one analyst remarked, 'The deserts of Libya have ... proved well able to absorb capital investment, but often with no tangible return.'[19]

The most positive alternative source of revenue for Libya, with its sunny climate, attractive shoreline and wealth of antiquities, is tourism. Given the political hostility towards Libya in the mid-1990s, however, tourists are unlikely to arrive in droves for a while.

Notes

1. It has been estimated that the Italian government spent more than £50 million on development projects in Libya over 30 years, mostly on agriculture and land reclamation, but the economy remained in default over this period. See Rawle Farley, *Planning for Development in Libya* (1971), p. 164.
2. Cited in John Wright, *Libya* (1969), p. 224.
3. These agencies included the Libyan Public Development and Stabilization Agency (LPDSA), a mainly British organization, and several US agencies, including the Libyan American Technical Assistance Service (LATAS), the Libyan American Joint Services (LAJS), and the Libyan American Reconstruction Agency (LARC).
4. Interview in *al-Ahram* (Cairo), as reported in *Middle East Economic Survey*, XIII/52, 23 October, 1970, p. 1.
5. Interview in *Le Monde*, reported in *Middle East Economic Survey*, XVI/20, 9 March, 1973, pp. 10–11.
6. Committee for Middle East Trade (COMET), 'Libya: The Five-Year Development Plan 1981–85' (1982), pp. 9–10.
7. Other companies involved in the construction and management of the complex included Balleli and Tecnimont from Italy and Monenco from Canada.
8. In 1980 it was estimated that wheat grown at Sarir, including its transport to coastal markets, cost $1350 per tonne, compared to a contemporary UK wheat market price of $241 per tonne. See 'Libya Ploughs its Wealth into Agriculture', *Middle East Economic Digest*, 5 June, 1981, p. 24.
9. British, German and Italian companies have built a number of desalination plants. Most are located near large municipal centres and industrial complexes and serve to augment domestic water supply. Smaller units have also been constructed for the treatment of brackish and salty groundwaters. See Omar M. Salem, 'The Great Man-Made River Project', *Water Resources Development*, 8/4, 1992, p. 271.
10. For a discussion of this argument, see J.A. Allan, 'Natural Resources: Not so Natural for Ease of Development', in M.M. Buru (ed.), *Libya: State and Region: A Study of Regional Evolution* (1989).
11. In 1992, there were an estimated two million foreigners in Libya, including one million Egyptian farmers. Some 600,000 workers from South Korea, the Philippines, Thailand and Vietnam were engaged in construction of the Great Man-Made River. The 40,000 Tunisians who were expelled in 1985 had returned.
12. See Richard E. Mattione, *OPEC's Investments and the International Financial System* (1985), p. 136.
13. See Traute Wohlers-Scharf, 'Arab and Islamic Banks' (1983).
14. In May 1994, the US Treasury's Office of Foreign Asset Control published a list of 97 financial institutions with Libyan banking

interests. See *Middle East Economic Survey*, 37/34, 23 May, 1994, pp. B1–B4 for this list.

15. See Charles Tripp, 'La Libye et L'Afrique', *Politique Étrangère*, 49/2, 1984, pp. 317–18.
16. See René Otayek, 'La Libye Révolutionnaire au Sud du Sahara', *Magreb Machrek*, 94, October–December 1981, p. 14.
17. COMET Report, op. cit., p. 9.
18. See COMET Report, p. 11.
19. J. A. Allan, *Libya: The Experience of Oil* (1981), p. 68.

11 WHAT THE FUTURE MAY BRING

As seen earlier, when Libya emerged as an independent country after the Second World War, it was desperately poor and unable to survive without massive foreign aid. The discovery of large quantities of excellent quality oil beneath its deserts changed this pathetic scene into one of remarkable new-found wealth and prosperity. The international admiration and charitable feelings which this produced turned to bitterness a decade later when a new government took an aggressive stand regarding its share of the profits of oil production. In the end, this government overplayed its hand, individually and as a member of OPEC, failing to recognize changing market circumstances and the consequences of taking a radical position in the international political arena. The Libyan oil industry faded from the spotlight and was largely ignored outside of continental Europe even though it continued to produce some 1.5 million b/d.

Will Libya continue to be the forgotten producer in the future, or will it once again come into focus as an important oil province? Ultimately this depends on the extent of its oil and gas reserves and on the oil policies of its government. With regard to the former, the potential certainly exists for the discovery of new fields and probably for increased output from existing fields. Many geologists believe that there are significant amounts of oil and gas yet to be found in the Murzuq and Ghadames Basins and that offshore fields have only started to be defined. Most doubt that there are undiscovered large fields in the Sirte Basin but agree that there are certainly smaller fields to be located. They base their assumption on the fact that in the early days companies drilled a number of wells which displayed oil but which they did not develop at that time for various reasons. Some had more reserves than they could comfortably handle in their large, easily produced fields and left Libya without ever reexamining their minor discoveries. Others were put off by technical development problems which recent technology will be more easily able to overcome – the Mabruk field is a good example of this, but there are undoubtedly others.

With regard to Libyan oil policies, changes in the near term seem unlikely. Exploration ranks high on the government's agenda and there is no reason to believe that it will cease to invite foreign companies in, or that these will fail to respond to its overtures. The results of the EPSA III round of acreage allocations, however, were disappointing and it is likely that the government will find that it needs either to improve the negotiable terms of its standard EPSA III contract or announce an EPSA IV round designed to attract more widespread exploration. In addition, there were signs at the beginning of 1996 that it might be willing to offer seismic option contracts – which several companies have sought – under which it would grant a company the right to do seismic exploration of a given acreage. If the company liked what it found, it would be offered an EPSA contract to develop the acreage; if it did not like the results, it would hand over its seismic data to NOC and leave.

There is also the question of what changes might occur in oil policies following a future change of government in Libya, whether this occurs in the natural course of events or unexpectedly. There are far too many unknowns to be able to predict the general policies of whichever administration succeeds the rule of Colonel Qaddafi, or the reactions of the outside world. But with regard specifically to oil, if history is any indication, the chances are that the main thrust of future Libyan oil policy will be similar to that of the past. As this study has shown, there has been remarkable continuity in policies regarding the development and functioning of the industry from the beginning when the country was ruled by a monarch through the years of the Revolutionary Command Council. This is due to the fact that geology and geography, not political bias, determine how this crucial industry works. Despite the highly nationalistic and often virulently anti-Western extremism of the Qaddafi government's political rhetoric, the Libyan oil industry has been conducted in an essentially conservative and orthodox manner, with the exception of a brief period in the early 1970s. It is difficult to see why any new government would change this approach. On the other hand, if American companies returned to Libya in the course of time, this could alter sufficiently the thrust of exploration and production activity within the country, and perhaps marketing abroad, to bring about changes in government policies.

Sanctions

The continuation of UN and US sanctions could have an increasingly significant effect on the Libyan oil industry in the long term. In order to assess what this might be, it is necessary to look at the cause, evolution and effect of boycotts and sanctions over the years. Appendix 11.1 shows that restrictions by the US government on trade with Libya began in the late 1970s as a result of a belief that the Libyan government was supporting international terrorism and subversion. In 1982, the USA banned the import of Libyan oil and increased its restrictions on exports; three years later it extended the import ban to include crude products. These measures did not seriously damage the Libyan oil industry as it was easily able to secure markets for its crude and the products it exported in western and eastern Europe.

In 1984, the killing of a British policewoman, Yvonne Fletcher, outside the Libyan Embassy in London led to a break in UK–Libyan diplomatic relations; again, this did not have a marked effect on the Libyan oil economy. However, in 1986 the situation became more serious when US air raids on Tripoli and Benghazi in April were followed in June by a US government order that all US companies and citizens must leave the country. For the next six years, economic sanctions against Libya were limited to those imposed by the USA and, to a more limited extent, the UK. In 1991, however, the UN Security Council took action in response to indictments in US and UK courts of two Libyan nationals for involvement in the bombing in December 1988 of Pan American Flight 103 over Lockerbie, Scotland. In November 1993, the UN passed Resolution 833 mandating an embargo on the export to Libya of certain equipment and components used in export and refining facilities (see Appendix 11.2). It did not extend its prohibition to banning imports of Libyan crude and products but it did adopt measures freezing Libyan financial assets abroad. This resolution was extended subsequently at three-month intervals and was still in effect in 1996, with sporadic reports of sanctions having been broken by contractors said to have supplied Libya with banned equipment and spare parts.

UN Resolution 833 variously affects Libyan upstream

operations. Probably the most important restriction is its ban on the selling or leasing of technology licences. This hampers the use of enhanced oil recovery procedures in older Libyan oilfields and could have a serious effect on production from these fields if continued indefinitely. There are indications that difficulty with reservoir pressure currently is preventing state-operated fields from increasing production and that output from these fields will decline in 1996 unless circumstances change. Perhaps as a warning on this, the oil ministry has announced that the chronically flat OPEC quotas make increased capacity an expensive luxury. The government will lose income if the oil industry continues merely to tick over at a low rate and does not expand. This could further reduce the amounts available for investment in the industry, so making production increases and exploration by NOC even less possible.

There is an impact on the upstream operations of NOC – and of foreign oil companies working in Libya – from the ban on sales to Libya of capital equipment such as turbines and generators, which arises partly from UN Resolution 833 but more importantly from US restrictions. The UK also forbids the export to Libya of equipment or supplies involved in oil production. NOC faces a problem repairing and replacing parts of equipment installed many years ago by American companies in their production operations. In some cases, appropriate parts or new equipment cannot be found to maintain well output or to carry out specialized processing.

The effect of the UN Resolution on exporting oil would appear to be minimal. It bans the sale to Libya of pumps and other paraphernalia used in the loading of crude oil and products for shipment abroad. But presumably Libya is not finding it difficult to overcome this prohibition as the equipment in question (such as anchor chains) is neither unusual nor sophisticated.

The UN Resolution carries more bite on downstream activities by forbidding the export to Libya of specific items needed in refining as well as of licences for refining processes. The items involved are described in terms of US American Petroleum Institute or American Society of Mechanical Engineers standards and it seems conceivable that substitutes exist which are not covered by the UN list, although these may not be of the highest standard. The blacklisting of catalytic reactors and prepared catalysts is important as the licences for constructing such reactors are held by American companies. The ban on Libya's acquisition of them means that the country will continue to be unable to upgrade its refineries to make more gasoline for internal consumption. This means that the government must continue to set aside foreign exchange badly needed elsewhere for the import of light distillates, and also that NOC will be forced to continue to waste its best quality crude to make gasoline.

The UN freeze of funds held abroad by Libyan public authorities or by 'any Libyan undertaking' was intended to restrict downstream or other financial investment in Europe by the Libyan government or NOC. This was made plain by the fact that the UN specifically exempted funds derived from the sale or supply of crude oil, products or natural gas if these were paid into separate bank accounts exclusively set up for this purpose. The initial effect of this restriction on existing Libyan downstream activities in Europe was forestalled by the Libyan government reducing its holdings in companies involved in these functions. In the mid-1990s it appeared that, although its equity stake had been reduced in such operations, Libya was continuing to play a significant role in the direction of these companies as a result of their dependence on quality Libyan crudes in their refineries. Presumably, this will continue.

Probably the most damaging effect of long-term sanctions would be the continued isolation of the Libyan oil industry from the mainstream of exploration and production operations in the rest of the world. Without exception, everyone in the industry, from policy makers in the ministry and managers in NOC down to technical personnel, will suffer increasingly from the loss of the collegiate benefits which come from membership in an evolving business. These include not only learning about technologies, especially those which are computer-based, but also training in the use of new production and refining techniques and equipment. On a broader level, what they are missing is the opportunity for a continuing exchange of ideas and problem-solving.

Threat of Increased Sanctions

In the early 1990s, the US government was pressing for more drastic measures to cripple the Libyan oil industry but found European governments opposed to this idea. At the end of 1995, the US Congress voted for severe US sanctions against both Iran and Libya in a single bill. The fate of this legislation was still undecided at the time that this study went to the printer. If it does become law, or if similar proposals become law in the future, the effect on several European companies involved in Libyan oil operations could be serious.

The pending legislation imposes mandatory sanctions on foreign companies that contribute to the development of petroleum resources in Libya and especially those that make new investments of more than $40 million in Libya's oil and gas sector. The punishments against such companies include restrictions on access to credit from US financial institutions including the Export-Import Bank and a ban on licences for the export to them of goods or technologies.

The motives behind this so-called 'secondary boycott' are political and complex, connected partly to the US presidential and congressional elections to be held in November 1996. Some of the companies which could be affected by such measures are competitors to US oil companies. The prime example is Total, which recently won an important contract for development of the Iranian Sirri oil and gas field which the US government had forced Conoco to give up, and which also recently won a ruling in France for the title to WR Grace's 12 per cent stake in the Libyan Mabruk field.[1] Total has refining capacity and retail outlets in the United States which it would have difficulty managing if it was faced with sanctions. Total will not be ready to let major engineering and equipment contracts for the Mabruk field – contracts which will be worth far more than $40 million – until 1997. The other company involved in a major Libyan operation in 1996, Repsol, has less to fear. It awarded the final contract for the main engineering work, including construction and management, for the Murzuq field early in 1996. It has also already awarded contracts for pipeline construction and equipment supplies.

The Other Side of the Coin

In some ways, the Libyan oil industry has become more robust because of sanctions and this is likely to continue. NOC takes considerable pride in being a survivor, in its ability to keep production going in the oilfields which the Americans left behind in 1986 and in being able to find secure markets for its exports which it prices in terms of the Brent market. Its long-term crude customers for 1996, for instance, are largely the same as those of the previous year and it is continuing to sell refined products in Europe despite the loss this year of the European Union's tax exemption for petroleum products from developing countries.

These not inconsequential successes have strengthened the resolve of both the oil ministry and NOC to continue the development of the Libyan oil industry on their terms. Foreign oil companies sometimes make the mistake of assuming that Libya is desperate and begging for their help. When they get to the negotiating table, they find it hard to believe that NOC will continue to insist on an investment of specific amounts of money, seismic and drilling in the acreage they are requesting, and that it is unwilling to bring in companies whose only interest is in development, not exploration. They do not realize that it will only consider companies which have paid their exploration dues and have a long track record in Libya, like Repsol, Total and OMV, for the task of bringing on stream already known, but undeveloped, reserves. They overlook the fact that NOC personnel are government employees and therefore do not have the incentive which employees of private companies have, of a promised bonus for completing a contract.

Sanctions have cemented stronger links, upstream and downstream, with continental European companies. The threatened additional US measures have strengthened further the ties between Europe and Libya, involving governments as well as companies. The European American Chamber of Commerce has argued that any attempt to bar imports into the United States by a targeted and sanctioned company would violate provisions of the World Trade Organization and invoke retaliation against US companies. The European Commission sent a letter of protest to the US State Department and the US Congress arguing that the

> United States has no basis in international law to claim the right to regulate in any way transactions taking place outside the United States. This way of unilaterally attempting to impose policies on third parties disturbs international trade and investment relations and depreciates the standing of internationally accepted fora for any such measures.[2]

The fact that the US majors are unable to enter Libya greatly increases its attraction for European and other non-American companies and especially for smaller companies which lack the financial resources to compete with the majors on the same playing field. Many of these consider Libya to be an essentially stable oil province with a national oil company which observes standard business practices and with a government which maintains good relations with foreign companies – both those with long-standing operations in the country and those which would like to sign exploration and production-sharing agreements. They are unlikely to change their views unless something totally unpredictable occurs.

Notes

1. Grace lost its claim on the grounds that its original partnership agreement was with the state-owned NOC rather than with the Libyan government. See *Oil Daily Energy Compass*, 'Compass Fax', 15 February, 1996.
2. For text of letter see *Middle East Economic Survey*, 39/20, 12 February, 1996, p. D2.

APPENDIX 11.1

CHRONOLOGY OF SANCTIONS AGAINST LIBYA

1978 May	USA bans military exports to Libya including aircraft and some agricultural and electronic equipment.
1981	USA requests all US citizens to leave Libya.
1982 March	USA bans import of Libyan oil and places restrictions on US exports to Libya. Cites alleged support for international terrorism and subversion.
1983	USA asks other nations to support curbs on exports to Libya.
1984 April	Policewoman Yvonne Fletcher shot and killed outside Libyan Embassy in London. UK breaks off diplomatic relations with Libya.
1985	USA bans import of Libyan crude products. Products refined in Caribbean from Libyan oil by Amerada Hess exempt.
1986 January	USA extends ban to cover all US import/export trade with Libya. Blocks Libyan funds in USA. Cites attacks on Rome and Vienna airports December 1985.
1986 April	USA conducts air raids on Tripoli and Benghazi. EC reduces Libyan diplomatic missions in Europe.
1986 June	USA orders all US oil companies to leave Libya.
1986 December	US ban on 'all direct economic activities between the US or US nationals with Libya' extended for one year.

1987	December	US ban on economic activities extended for another year.
1988	December	US ban on economic activities extended until June 1989.
1988	December	Pan Am flight 103 blown up over Lockerbie in Scotland.
1989	July	US ban on economic activities extended to December 1989.
1989	September	French airliner crashes in Niger.
1990	January	US ban on economic activities extended another year.
1991	January	US ban on economic activities extended another year.
1991	August	US trade embargo extended to Libyan downstream operations in Europe, including Holborn and Oilinvest.
1991	October	French magistrate charges several Libyans with blowing up the French airliner which crashed in Niger.
1991	October	Indictments in US and UK courts of two Libyan nationals, Abdelhasset al-Maghrahi and Al-Amin Khalifah al-Fahimah, for Lockerbie aircraft bombing.
1992	January	UN Security Council Resolution 731 urges Libya to respond to requests for cooperation in establishing responsibility for Lockerbie bombing.
1992	January	US ban on economic activities extended another year.
1992	March	UN Security Council Resolution 748 adopts sanctions for three months if Libya fails to surrender Lockerbie suspects. Prohibition of aircraft flights and weapons sales. Reduction

		of Libyan diplomatic missions abroad. USA adds more companies to its embargo list.
1992	September	UN sanctions renewed for three months.
1992	December	US ban on economic activities extended for another year. UN sanctions renewed for three months.
1993	February	USA bans US law firms or their foreign branches from offering legal services to the Libyan government or its agencies.
1993	April	UN sanctions renewed for another three months. 1 October deadline for Libya to hand over Lockerbie suspects for trial in USA or UK. USA trying to organize a global oil embargo.
1993	November	UN Resolution 833 mandates stricter embargo. Includes ban on exporting equipment and components for oil export and refining facilities to Libya but does not ban imports of Libyan oil and products. Also freezes Libyan financial assets abroad and restricts domestic airline operations.
1993	December	US ban on economic activities extended another year.
1994	February	USA calls for international embargo against Libyan oil trade. Extends embargo to Libyan-connected financial organizations.
1994		UN extends sanctions every three months throughout the year.
1995	April	UN extends sanctions every three months throughout the year. USA trying to get UN sanctions to include trade in oil.
1995	December	US ban on economic activities extended another year.

APPENDIX 11.2

UN SECURITY COUNCIL RESOLUTION 833 ADOPTED 11 NOVEMBER, 1993

Embargoed Items

I. Pumps of medium or large capacity whose capacity is equal to or larger than 350 cubic metres per hour and drivers (gas turbines and electric motors) designed for use in the transportation of crude oil and natural gas.

II. Equipment designed for use in crude oil export terminals:

- Loading buoys or single point moorings (spm).
- Flexible hoses for connection between underwater manifolds (plem) and single point mooring and floating loading hoses of large sizes (from 12 to 16 inches).
- Anchor chains.

III. Equipment not specially designed for use in crude oil export terminals but which because of their large capacity can be used for this purpose:

- Loading pumps of large capacity (4000 m^3/h) and small head (10 bars).
- Boosting pumps within the same range of flow rates.
- Inline pipeline inspection tools and cleaning devices (i.e. pigging tools) (16 inches and above).
- Metering equipment of large capacity (1000 m^3/h and above).

IV. Refinery equipment:

- Boilers meeting American Society of Mechanical Engineers 1 standards.
- Furnaces meeting American Society of Mechanical Engineers 8 standards.
- Fractionation columns meeting American Society of Mechanical Engineers 8 standards.
- Pumps meeting American Petroleum Institute 610 standards.
- Catalytic reactors meeting American Society of Mechanical Engineers 8 standards.
- Prepared catalysts, including the following:
  Catalysts containing platinum;
  Catalysts containing molybdenum.
- Spare parts destined for the items in I to IV above.

BIBLIOGRAPHY

Books and Journal Articles

Adelman, M.A. (1972), *The World Petroleum Market*, Baltimore, MD: Johns Hopkins University Press.

Algerian Government (1971), *Background Information on the Relationship between Algeria and the French Oil Companies*, Algiers: Government Press.

Allan, J.A. (1981), *Libya: The Experience of Oil*, London: Croom Helm.

Anastassopoulos, Jean-Pierre, Georges Blanc and Pierre Dussauge (1987), *State-Owned Multinationals*, New York: John Wiley.

Arab Banking Corporation (1994), *The Arab Economies: Structure and Outlook*, 4th Revised Edition, Manama: ABC, pp. 71–6.

Arab Petroleum Investments Corporation (1994), *The Investment Requirements for the Development of the Arab Oil and Gas Sector and Related Downstream Activities until the End of this Decade*, Cairo: APIC.

Arab Petroleum Research Centre (1994), *Arab Oil & Gas Directory*, Paris: APRS, pp. 229–66.

Ball, Max W. and Douglas Ball (1965), *This Fascinating Oil Business*, Indianapolis: Bobbs-Merrill.

Ball, Max W. and Douglas Ball (1953), 'Oil Prospects of Israel', *American Association of Petroleum Geologists Bulletin*, 37/1, pp. 1–113.

Barclay's (1994), 'Abecor Country Report: The Maghreb', London: Barclay's Bank.

Barker, P. and K.S. McLachlan (1982), 'Development of the Libyan Oil Industry' in J.A. Allan, (ed.), *Libya Since Independence*, London: Croom Helm, pp. 37–54.

Barrows, Gordon H. (1983), *Worldwide Concession Contracts and Petroleum Legislation*, Tulsa: PennWell, pp. 9–13, 16, 172–3.

Bellini, E. and D. Massa (1980), 'A Stratigraphic Contribution to the Palaeozoic of the Southern Basins of Libya' in M.J. Salem and M.T. Busrewil (eds), *The Geology of Libya: Second Symposium on the Geology of Libya held at Tripoli, 16–21 September, 1978*, vol. I, New York: Academic Press.

Benfield, A.C. and E.P. Wright (1980), 'Post-Eocene Sedimentation in the Eastern Sirt Basin, Libya' in *The Geology of Libya, op. cit.*, vol II, pp. 463–97.

Blair, John M. (1976), *The Control of Oil*, New York: Pantheon Books.

British Petroleum (1977), *Our Industry Petroleum*, London: British Petroleum.

Buckman, David (1991), 'Drive to Boost Recovery', *Petroleum Economist*, 58/3, pp. 6–7.

Buckman, David (1994), 'North Africa Sees Europe as Big Piped Gas Market', *Petroleum Review*, 48/574, pp. 523–6.

Buru, M.M. (ed.) (1989), *Libya: State and Region: A Study of Regional Evolution*, London: University of London Centre of Near and Middle East Studies.

Calvocoressi, Peter (1953), *Survey of International Affairs 1949–50*, Oxford: Oxford University Press, pp. 539–45.

Chadwick, Margaret, David Long and Machiko Nissanke (1987), *Soviet Oil Exports: Trade Adjustments, Refining Constraints and Market Behaviour*, Oxford: Oxford University Press.

Chester, Edward W. (1983), *United States Oil Policy and Diplomacy: A Twentieth-Century Overview*, Westport, CT: Greenwood Press.

Clarke, John I. (1963), 'Oil in Libya: Some Implications', *Economic Geography*, 39/1, pp. 40–59.

Committee for Middle East Trade (COMET) (1985), 'Libya: The Five-Year Development Plan 1981–1985', London: Comet.

Conant, Louis C. and Gus H. Goudarzi (1967), 'Stratigraphic and Tectonic Framework of Libya', *Bulletin of the American Association of Petroleum Geologists*, 61/5, pp. 5–16.

Danielsen, Albert L. (1982), *The Evolution of OPEC*, New York: Harcourt Brace Jovanovich.

Economist Intelligence Unit (1994), *EIU Country Profile 1994–95 Libya*, London: EIU.

Energy Information Administration, Department of Energy (1984), *The Petroleum Resources of Libya, Algeria, and Egypt*, Washington, DC: US Government Printing Office DOE/EIA-0435.

Engler, Robert (1961), *The Politics of Oil: A Study of Private Power and Democratic Directions*, New York: Macmillan.

Europa, *International Who's Who (1989)*, London: Europa, 53rd Edition.

Exploration Staff of the Arabian Gulf Oil Company (1980), 'Geology of a Stratigraphic Giant – the Messlah Oil Field' in *The Geology of Libya, op. cit.*, vol II, pp. 521–36.

Farley, Rawle (1971), *Planning for Development in Libya*, New York: Praeger.

Fesharaki, Fereidun and David Isaak (1984), *OPEC and the World Refining Crisis*, EIU Special Report 168, London: Economist Intelligence Unit.

Fried, Edward R. and Charles L. Schultze (eds) (1975), *Higher Oil Prices and the World Economy: The Adjustment Problem*, Washington, DC: Brookings Institution.

'Full Text of Law Establishing Libya's National Oil Company' (1968), *Middle East Economic Survey Supplement*, XI/26, pp. 1–9.

Ghanem, Shukri Mohammed (1975), *The Pricing of Libyan Oil*, Valletta, Malta: Adams Publishing House.

Gillespie, Kate and Clement Moor Henry (1995), *Oil in the New World Order*, Gainesville, FL: University Press of Florida.

Gohrbandt, Klaus H.A. (1967), 'Upper Cretaceous and Lower Tertiary Stratigraphy along the Western and Southwestern Edge of the Sirte Basin, Libya' in J. Williams (ed.), *South-Central Libya and Northern Chad*, Petroleum Exploration Society of Libya, Amsterdam: Drukherij Holland.

Goudarzi, Gus, (1980), 'Structure – Libya', in *The Geology of Libya, op. cit.*, vol III, p. 879–91.

Hammuda, Omar S. (1980), 'Geologic Factors Controlling Fluid Trapping and Anomalous Freshwater Occurrence in the Tadrat Sandstone, Al Hamahah al Hamra Area, Ghadamis Basin' in *The Geology of Libya, op.cit.*, vol. II, pp. 501–2.

Harris, Lilian Craig (1986), *Libya: Qadhafi's Revolution and the Modern State*, Boulder: Westview Press.

Hartshorn, J.E. (1978), *Objectives of the Petroleum Exporting Countries*, Nicosia, Cyprus: Middle East Petroleum and Economic Publications.

Hartshorn, J. E. (1962), *Oil Companies and Governments*, London: Faber and Faber, pp. 175–91.

Hepple, P. (ed.) (1969), *The Exploration for Petroleum in Europe and North Africa*, London: Institute of Petroleum.

Higgins, Benjamin (1959), *Economic Development, Principles, Problems and Policies*, New York: W.W. Norton.

Horsnell, Paul and Robert Mabro (1993), *Oil Markets and Prices: The Brent Market and the Formation of World Oil Prices*, Oxford: Oxford University Press.

International Bank for Reconstruction and Development (1960), *The Economic Development of Libya*, Baltimore, MD: Johns Hopkins.

Iskander, Marwan (1969), 'Economic Development Plans in Oil Exporting Countries and their Implications for Oil Production Targets', *Middle East Economic Survey*, XIII/19, Supplement.

Jarjour, Gabi (1982), *OPEC and Oil Pricing Structure: Analysis of OPEC Official Resolutions and Press Releases 1962–1982*, Dhahran, Saudi Arabia: University of Petroleum and Minerals.

Jenkins, Gilbert (1984), *Oil Economists' Handbook*, London: Applied Science Publishers, p. 116.

Jensen, W.G. (1967), *Energy in Europe 1945–1980*, London: G.T. Foulis & Co., pp. 58–61.

Kennedy, John L. (1994), 'OMV to Focus on Restructuring, Integrated Oil Operations', *Oil & Gas Journal*, 92/15, pp. 32–4.

Kent, P.E., (1969), 'The Geological Framework of Petroleum Exploration in Europe and North Africa and the Implications of Continental Drift Hypotheses' in P. Hepple, *The Exploration for Petroleum in Europe and North Africa*, London: Institute of Petroleum, pp. 3–25.

Kingdom of Libya, Ministry of Petroleum Affairs (1968), *Libyan Oil 1954–1967*, Tripoli.

Kogbe, Cornelius A. (1980), 'The Trans-Saharan Seaway During the Cretaceous' in *The Geology of Libya, op. cit.*, vol. I, pp. 91–2.

Kosloff, I.R. (1953), 'Oil Development in Israel: Introduction', *Economic News* (Tel Aviv) vol. V, p. iii.

Kubbah, Abdul A.Q. (1964), *Libya: Its Oil Industry and Economic System*, Beirut: Rihani Press.

Larson, Henrietta M. *et. al.*, (1971), *History of Standard Oil Company (New Jersey): New Horizons 1927–1950*, New York: Harper & Row.

'Lasmo Sings Libya E&P Deal' (1990), *Platt's Oilgram News*, 68/200, p. 2.

Lehrman, Hal, (1954), 'The Turks like American Capitalists', *Fortune*, 24/12, p. 122.

Lenczowski, George (1960), *Oil and State in the Middle East*, Ithaca: Cornell University Press.

'Libya Ploughs its Wealth into Agriculture'(1981), *Middle East Economic Digest*, 25/23, pp. 24–6.

'Libya Seeks Refinery Products Customers', *Lloyds' List*, 7 October, 1983.

'Libya to Earn Stake in German Refinery', *Oil & Gas Journal*, 86/21, p. 16.

Libyan National Oil Company (1988), 'The Oil Industry in the Great Jamahiriya', Tripoli: Government Printing Office.

Longrigg, Stephen H. (1961), *Oil in the Middle East*, Second Edition, Oxford: Oxford University Press.

Lorenz, John (1980), 'Late Jurassic–Early Cretaceous Sedimentation and Tectonics of the Murzuk Basin, Southwestern Libya' in *The Geology of Libya*, *op. cit.*, vol II, pp. 383–5.

Luciani, Giacomo (1984), *The Oil Companies and the Arab World*, London: Croom Helm.

Mabro, Robert (ed.) (1986), *OPEC and the World Oil Market: The Genesis of the 1986 Price Crisis*, Oxford: Oxford University Press.

Mansfield, David (1978), 'Check to Production Buildup', *Petroleum Economist*, XLV/9, pp. 371–3.

Mattione, Richard E. (1985), *OPEC's Investments and the International Financial System*, Washington, DC: Brookings Institution.

McDonald, Stephen L. (1963), 'Federal Tax Treatment of Income from Oil and Gas', Washington, D.C. : Brookings Institution.

Ministry of Petroleum Affairs (1965), 'Petroleum Development in Libya 1954 through 1964', Tripoli: Government Press.

'Model Exploration and Production Sharing Agreement of 1990' (1992), *Barrows Petroleum Legislation Supplement 109*, Appendix II, New York: Barrows.

Molle, Willem and Egbert Wever (1984), *Oil Refineries and Petrochemical Industries in Western Europe*, Aldershot: Gower.

Mommer, Bernard (1988), *La Cuestion Petrolera*, Caracas: Fondo Editorial Tropilos.

Morgan, Trevor (1991), 'The Economics of Sulphur in Heavy Fuel Oil', *Petroleum Review*, 45/528, pp. 36–8.

'Moving Liquefied Natural Gas from Libya to Italy and Spain'(1965), *Esso Magazine*.

Nehring, Richard (1978), *Giant Oil Fields and World Oil Resources*, Santa Monica: Rand, R-2284-CIA, pp.136–7.

'New Pipeline and Oil Terminal Boost Libya's Oil Exports'(1967), *Europe & Oil*.

Nyrop, Richard F. *et. al.* (1973), *Area Handbook for Libya*, Washington, DC: US Government Printing Office.

Organization of Arab Petroleum Exporting Countries (1990), *Prospects of Arab Petroleum Refining Industry: Handbook of Arab Oil Refineries*, Kuwait: OAPEC.

Otayek, René (1981), 'La Libye Révolutionnaire au Sud du Sahara', *Magreb Machrek*, 94, p. 5–20.

Petroleum Commission (1960), 'Petroleum Development in Libya 1954 through mid 1960', Tripoli: Government Press.

Petroleum Commission (1961), 'Petroleum Development in Libya 1954 through mid 1961', Tripoli: Government Press.

Petroleum Commission (1962), 'Petroleum Development in Libya 1954 through mid 1962', Tripoli: Government Press.

Petroleum Economist and Kennet Oil Logistics (1995), *Oil Logistics Guide to Northern Europe and the Mediterranean 1995*, London: Petroleum Economist, pp. 146, 157, 171.

Petrostrategies (1978), *Soviet Oil, Gas and Energy Databook*, Stavanger: Noroil Publishing House.

Redden, Kenneth and Jon Huston (1956), *The Petroleum Law of Turkey*, Istanbul: Fakulteler Matbaasi.

Reyment, R.A. and E.R. Reyment (1980), 'The Palaeocene Trans-Saharan Transgression and its Ostracod Fauna' in *The Geology of Libya*, *op. cit.*, vol. I, pp 245–7.

Russell, Jeremy (1978), *Energy as a Factor in Soviet Foreign Policy*, Farnborough, Hants: Saxon House.

Salem, M.J. and M.T. Busrewil (eds) (1980), *The Geology of Libya*, Second Symposium on the Geology of Libya held at Tripoli, 16–21 September, 1978, New York: Academic Press.

Salem, Omar M. (1992), 'The Great Man-Made River Project', *Water Resources Development*, 8/4, pp. 270–8.

Salomon Brothers (1995), 'OMV: Buy: Accelerating Pace of Recovery', *European Equity Research: European Oils*, New York: Salomon Brothers.

Schlosberg, M.D. and L. Kuenstler (1953), 'The Israel Petroleum Law 5712-1952 and Petroleum Regulations 5713-1953', *Economic News* (Tel Aviv), vol. V, pp. 3–7.

Segal, Jeffrey (1979), 'Future Clouded by Uncertainty', *Petroleum Economist World Survey LNG Market*, XLVI/12, pp. 515–20.

Segal, Jeffrey (1980), 'Slower Growth for the 1980s', *Petroleum Economist World Survey LNG Market*, XLVII/12, pp. 513–15.

Segal, Jeffrey (1981), 'Pricing Structure in Disarray', *Petroleum Economist World Survey LNG Market*, XLVIII/12, pp. 517–20.

Selley, Richard C. (1985), *Elements of Petroleum Geology*, New York: W.H. Freeman.

Seymour, Ian (1980), *OPEC Instrument of Change*, London: Macmillan.

Shwadran, Benjamin (1985), *The Middle East, Oil and the Great Powers*, Boulder: Westview Press.

Simons, Geoff (1993), *Libya: The Struggle for Survival*, London: Macmillan.

Skeet, Ian (1988), *OPEC: Twenty-Five Years of Prices and Politics*, Cambridge: Cambridge University Press.

Tamoil (1992), 'Oilinvest in Europe: A Simple Idea, A Big Project', *Tamoil Magazine*, February, Italy: Tamoil.

Terry, C.E. and J.J. Williams (1969), 'The Idris 'A' Bioherm and Oilfield, Sirte Basin, Libya – Its Commercial Development, Regional Palaeocene Geologic Setting and Stratigraphy' in Hepple, *op. cit.*, pp. 31–48.

Tetreault, Mary Ann (1985), *Revolution in the World Petroleum Market*, Westport: Quorum Books.

Thomas, David (1995), 'Exploration Limited Since 70s in Libya's Sirte Basin', *Oil & Gas Journal*, 93/11, pp. 99–104.

Thomas, David (1995), 'Geology, Murzuk Oil Development Could Boost S.W. Libya Prospects', *Oil & Gas Journal*, 93/10, pp. 41–6.

Tiratsoo, E.N. (1984), *Oilfields of the World*, Beaconsfield: Scientific Press, pp. 216–9.

Tripp, Charles (1984), 'La Libye et L'Afrique', *Politique Étrangère*, 49/2, pp. 317–22.

Turner, Brian R. (1980), 'Palaeozoic Sedimentology of the Southeastern Part of Al Kufrah Basin, Libya' in *The Geology of Libya, op. cit.*, vol II, pp. 351–72.

United Arab Republic (1960), *Petroleum in United Arab Republic*, Cairo: Government Press.

United Nations (1955), *Supplement to World Economic Report*, 'Summary of Recent Economic Developments in Africa 1952–53', E/2522/S/ECA/26, New York: UN.

United Nations Development Programme (1995), *Human Development Report 1995*, Oxford: Oxford University Press.

van der Linde, Coby (1991), *Dynamic International Oil Markets: Oil Market Developments and Structure 1960–1990*, Dordnecht: Kluwer Academic Publishers.

Villard, Henry S. (1956), *Libya: The New Arab Kingdom of North Africa*, Ithaca, NY: Cornell University Press.

Waddams, Frank C. (1980), *The Libyan Oil Industry*, London: Croom Helm.

Wall, Bernard (1988), *Growth in a Changing Environment: A History of Standard Oil Company (New Jersey) 1950–1972 and Exxon Corporation 1972–1975*, New York: McGraw-Hill.

Williams, Bob (1988), 'OPEC Ventures Downstream: Industry Threat or Stability Aid?', *Oil & Gas Journal*, 86/20, 16 May, pp. 15–17.

Wilson, David Cameron (1991), *CIS and East European Databook*, London: Eastern Bloc Research Ltd.

Wilson, David Cameron (1995), *CIS and East European Databook*, London: Eastern Bloc Research Ltd.

Wohlers-Scharf, Traute (1983), 'Arab and Islamic Banks', Paris: OECD Development Centre.

World Bank Mission (1960), *The Economic Development of Libya*, Baltimore, MD: Johns Hopkins Press.

Wright, John (1969), *Libya*, London: Ernest Benn.

Wright, John (1981), *Libya: A Modern History*, London: Croom Helm.

Annuals, Journals and Newsletters

An-Nahar Arab Report & Memo
Arab Report & Record
Barrows Petroleum Legislation
BP Statistical Review of World Energy
Cedigaz Natural Gas in the World Survey
Cedigaz News Report
Energy Compass
European Chemical News
Financial Times International Yearbooks Oil & Gas
International Energy Agency Oil and Gas Information
International Crude Oil and Product Prices
International Legal Materials Current Documents
International Petroleum Encyclopedia
Martindale-Hubbell Law Directory
Middle East Economic Digest
Middle East Economic Survey
Middle East Report
Mideast Mirror
Oil & Gas Journal
Oil & Gas Journal Data Book
OPEC Annual Statistical Bulletin
Petroleum Economist
Petroleum Intelligence Weekly
Petroleum Intelligence Weekly International Crude Oil Market Handbook
Petroleum Press Service
Platt's Oil Price Handbook and Oilmanac
Platt's Oilgram Price Report
United Nations Annual Bulletin of Gas Statistics for Europe
Weekly Petroleum Argus
World Oil

Company Annual Reports

Agip
Chieftain International, Inc.
Hardy Oil & Gas Plc
Husky Oil Ltd
International Petroleum Corporation
Lasmo Plc
Nova Corporation
OAPEC
OMV

OPEC
PanCanadian Petroleum Limited
Pedco
Petrobras
Petrofina
Repsol
Royal Dutch/Shell Group of Companies
Saga Petroleum
Société Nationale Elf Aquitaine
Standard Oil of New Jersey (SONJ)
Tamoil
Total
Veba
Veba Oel
Westcoast Energy Inc.
Wintershall

Archives

BP Archives at the University of Warwick.

INDEX

3 1542 00178 8532